# Rule-of-Thumb
# Cost Estimating
# for Building
# Mechanical Systems

# Rule-of-Thumb Cost Estimating for Building Mechanical Systems

**Accurate Estimating and Budgeting Using Unit Assembly Costs**

James H. Konkel, P.E., C.C.E.

**McGraw-Hill Book Company**

New York   St. Louis   San Francisco   Auckland   Bogotá
Hamburg   London   Madrid   Mexico
Milan   Montreal   New Delhi   Panama
Paris   São Paulo   Singapore
Sydney   Tokyo   Toronto

Library of Congress Cataloging-in-Publication Data

Konkel, James H.
  Rule-of-thumb cost estimating for building
mechanical systems.

  Bibliography: p.
  Includes index.
  1. Heating—Estimates. 2. Air conditioning—
Estimates. 3. Ventilation—Estimates. 4. Fire
detectors—Estimates. I. Title.
TH7335.K56 1987 | 697      86-27565
ISBN 0-07-044957-0

88-B1694

1234567890   DOC/DOC   893210987

ISBN 0-07-044957-0

*The editors for this book were Betty Sun and Dennis Gleason,
the designer was Naomi Auerbach, and the production
supervisor was Tom Kowalczyk. It was set in Century Schoolbook
by William Byrd Press, Inc.*

*Printed and bound by R. R. Donnelley & Sons Company*

# Contents

Preface     vii

1. Principles Involved in This Method                                    1

2. General Support Material                                              5

3. How to Start the Estimate                                           19

4. Sample Estimate                                                     22

5. Site Work                                                           51

6. Plumbing                                                            55

7. Heating and Cooling                                                 68

8. Packaged Heating, Cooling, and HVAC Units                         134

9. Air Distribution Equipment                                        147

10. Controls and Instrumentation                                     167

11. Balancing                                                        190

12. Insulation                                                       198

13. Fire Protection                                                  202

14. Special Systems                                                  210

15. Cost Estimating Summary                                          215


Bibliography     219
Appendix A Table of Equivalents     220
Appendix B Mechanical Legend     221
Index     223

# Preface

Two types of cost estimates have been employed in the past and are in common use today. The first type of estimate has usually been employed in the conceptual and early stages of a project. In the area of mechanical systems in buildings, typical examples of these early estimates are cost per square foot of building floor area, cost per ton of air conditioning load, and cost per fixture for plumbing.

These single-number estimates are based on previous job costs. They have proven to be risky at best and can lead one to proceed further into the design of the project than might be wise, only in order to obtain the next type of estimate currently in use today, the "labor and material take-off" estimate.

A labor and material take-off estimate can be done at any stage of preparation of plans and specifications. Historically, however, they have not been employed to any great degree until plans and specifications have been completed. When the job has been estimated from preliminary or incomplete plans by the "take-off" method, these estimates have proven to be inaccurate and typically have been below the actual final costs.

What has been needed, then, in the early stages of a project and before beginning detailed design, is a cost estimating method more accurate than the first and not as expensive and time consuming as the second. Over the past 25 years, the firm of McFall-Konkel & Kimball has developed such a method and proven it accurate. In 1980–81 we refined and extended the method. Our cost estimating results continue to be excellent.

The method uses labor and material costs from past jobs to form the basis for an installed cost estimate for a new project. Rather than having to price each piece of equipment, material, and labor individually for the new project, this system has grouped certain portions of the project into what might be called *unit operations* and their components, or *unit assemblies*. All of the components within each unit operation are itemized, priced, and plotted by size of the unit operation. An example of a unit operation would be a boiler. Every water boiler operation includes a boiler, burner, combustion air, flue, shutoff valves, certain piping, fuel supply, expansion tank, water makeup valves, and air elimination system. The size of each component varies with the size of the boiler. It can be seen, then, that the labor and material costs for each size of boiler and each of its components can be priced and plotted to form a curve representing the installed dollar cost per Btu/hr for any boiler operation, thus making it possible for the estimator to find the total installed cost of the boiler operation for the project simply by knowing its size.

Using this precept for each unit operation or unit assembly in the mechanical system, a complete cost estimating method has been produced and included in this book. From preliminary architectural plans, the designer, estimator, owner, or agency can accurately estimate the mechanical costs with very little design time involved.

We expect that this book will be an invaluable aid to architects, consulting engineers, building committees, governmental agencies, physical plant personnel, lenders, developers, mechanical contractors, and all others who need reliable costs of the mechanical systems in a building project early in the game. The book also provides a convenient method to easily compare the costs of different mechanical systems or components to help the designer decide on which system or component would be the best for the building. Thus it is an invaluable tool for the engineer in determining the most cost efficient system and its components.

The years of experience gained by the author and the staff of McFall-Konkel & Kimball, Inc., resulted largely from the research and development done by HVAC equipment manufacturers and their input to ASHRAE. Their design manuals, engineering bulletins, product manuals, and pricing manuals hold a wealth of information which the author was fortunate to have for his use over the years and in writing this book. Many manufacturers' names are mentioned in the book; they or their representatives gave freely of their time to this effort.

A number of manufacturers' representatives in Denver helped with pricing, among them were the CBA Co., CFM Co., Haynes/Trane Co., and Long & Associates. Special thanks to R. P. (Bob) Koenig of Autocon, Inc., who helped with the control section, and Gary Griffith of Griffith Engineering Service who helped with the balancing section.

Thanks also to the officers and staff of McFall-Konkel & Kimball, to Scott Lohr in our office who wrote the chapter on Fire Protection, to Bob Staskiewicz and Karen Jo Swieso who did the drafting, and to Kathy Leach who typed all of the manuscript in the book.

*James H. Konkel, P.E.*

# 1

# Principles Involved

## Background

When we first started to put this book together for our office in 1980, one of our primary goals was to assemble the material in such a way that even the newest and least-experienced engineer could make as good an estimate as the senior project engineers. However, once we listed the principles on which the book was to be based, it became apparent that experience and judgment could not be replaced. While we feel that a cost estimate of reasonable accuracy can be arrived at by inexperienced people using this book, a person who is experienced and astute in the marketplace will probably make a better estimate than the beginner a large percentage of the time.

## General Principles of Estimating

All estimating methods are based on certain principles. As indicated before, the more familiar a person is with these principles and the more adroit one is at working with them, the better the estimate will be. Principles involved in using this book will include the following:

1. The project estimate must be done using values for systems that may be easily verified from the final contract costs. Two things come to mind. The estimate must be arranged by trade, contract, and subcontract so that it matches (as nearly as possible) the contractor's method of assembling costs. Thus, a comparison of estimate vs. actual cost can be made and fed back into the system for future work. Also, when graphing the cost of a unit operation, say a heating boiler plant, dollars per Btu of capacity can be verified much more easily than can a cost expressed as dollars per square foot of building area.

2. A good cost estimate must assure that every operation and cost

is accounted for by a line item so that something is not forgotten and left out. We know that some costs will be estimated higher than the actual cost and some will be estimated lower than the actual cost. Statistically we know that the more line items we can include in the estimate the more nearly correct the total will be.

In the early stages of a project we are more interested in the accuracy of the total cost than the accuracy of each item.

3. Costs vary by project location. Different cost indices have been compiled by various groups for most major cities in the country. The information in this book was based on Denver, Colorado, costs. The 1986 *Dodge Manual for Building Construction Pricing and Scheduling* lists Denver's general adjustment index as 1.02. The estimators should adjust their final costs to their particular location. This can be done as follows:

$$\text{This book's cost} \times \frac{\text{your area index}}{\text{Denver index}} = \text{your area cost}$$

4. All prices should be based on costs at some date in time. Since it is nearly impossible to price every piece of material used in any book on the exact same date, we have tried to assemble all of the different actual costs and relate them to a common date of January 1, 1986. Using Hensel Phelps Construction Company's "Colorado Building Construction Cost Increase Averages" chart, we have projected these costs to January 1, 1987. This increase appears to be in the vicinity of only 2%.

In addition, when preparing an early estimate of the cost of a project, the estimator must make a judgment on the date of completion of the project and apply an inflation factor to cover the cost increase expected through this period. Obviously the higher the annual inflation rate, the more difficult it is to predict labor contract cost increases and material cost increases.

It appears to the author that an annual cost increase for building construction for 1987 will be in the area of 3%, and that by applying a 3% per year increase to the estimate, from the time the estimate is made to the midpoint of a construction, might be reasonable except in isolated situations.

5. The cost of a project is affected by the general conditions under which the contract is to be administered. Costs in this manual are based on our standard specifications, which are written to meet the Denver area standard of performance. More stringent specifications and general conditions cause the cost to rise. Some examples would be

*a.* Time-constricted construction period.

*b.* Completion date with heavy penalties for noncompliance.

*c.* Excessive warranty and guarantee requirements which are over and above the normal 1-year guarantee of workmanship and material.

*d.* Bid bond and performance bond costs must be added to costs of the contract.

6. The cost of the project will also be affected by unusual economic conditions at the time of bidding or expected during the life of the construction. This item should be assessed when making an estimate.

7. The cost will be affected by weather conditions during the life of the construction. Any anticipated unusually harsh weather should be noted and factored into labor costs. The Visitor's Center at the top of Pike's Peak, for which we designed the mechanical system a number of years ago, is an excellent example. The altitude where the building was to be built was 14,100 ft above sea level. Not only were temperatures extremely cold, winds of 60 to 100 mph were anticipated to prevail continually during the construction period. Another factor not normally encountered came into play. At 14,000 ft elevation, the ability of people to perform both physical and mental tasks is greatly diminished and labor production is measurably reduced. This phenomenon brings us to number 8 in our list of principles.

8. The capacity of some pieces of heating and cooling equipment is reduced when that equipment is employed at higher altitudes. While this book is not a design manual and will make little attempt to instruct the estimator on how and when to derate equipment capacity for altitude, the estimator must be competent to adjust equipment sizes and costs when the project is located at any different elevation.

For the purposes of this book, the costs are good for elevations up to 2000 ft above sea level. At altitudes above this, the affected equipment capacities should be derated at approximately 4% per 1,000 ft of elevation above sea level and the larger equipment chosen and priced for the project.

9. Costs of some building components, including heating, ventilating, and air conditioning (HVAC) systems, vary with outdoor weather conditions or so-called design temperatures. Mild winter temperatures result in smaller heating systems being installed and vice versa. Cooling systems vary in size not only with the temperature differential from outdoor to indoor temperatures but also with humidity conditions.

Denver has a very dry climate, and practically speaking, 100% of the cooling load is sensible load. Other areas in the country have a sizable percentage of the load in latent heat. Offsetting this smaller chiller load to some extent is Denver's high altitude, where the air has only 83% of the density of the air at sea level. Consequently the capacity of heating equipment with atmospheric-type burners, gravity heating equipment, and air-moving equipment must be derated.

These factors have not been adequately addressed in typical rule-of-thumb estimate numbers in the past. This estimating system automatically accounts for the design conditions because the estimator chooses each component price from its size, which was arrived at by design for the project.

10. Travel time to and from the project affects the project cost. If the project location takes excessive time for the labor force to reach and return daily, a factor for this cost must be included.

11. Task difficulty can change normal costs either up or down. For equipment and material installed at 20 ft above floor level, the costs are more than if that equipment were installed at 10 ft above floor level. On the other hand, tasks that become repetitive enough to decrease learning time will lower the cost.

12. Costs used in this book are based on union labor rates. Open shop costs may vary from these.

13. The *fundamental principle* to this book is that the unit cost diminishes as the size increases. Using this fundamental principle we can see that

- The unit cost of manufacturing equipment decreases with an increase in equipment size.

- The unit cost of installation of equipment decreases with an increase in equipment size.

- The unit cost of sales decreases as the size of equipment increases and as the size of the project increases.

- The unit cost of overhead for the contractor decreases as the project size increases. The contractor's need for actual dollar profit when expressed as a percentage of the total cost also decreases as the project size increases.

The essence of this book, then, is that each operation in a mechanical system can be plotted as cost relative to size and that size of the project is a controlling factor in estimating the cost of that project.

# 2

# General Support Material

## Labor Rates

Labor rates used in this book were derived from the labor rates shown in the 1986 *Dodge Manual for Building Construction Pricing and Scheduling*, adjusted to arrive at the following:

| | |
|---|---|
| Pipe fitter | $30/hr |
| Plumber | $30/hr |
| Sheet metal worker | $30/hr |
| Insulator | $27/hr |
| Balancing | $33/hr |
| Control fitter | $30/hr |
| Laborer-helper | $20/hr |

These rates include the base rate plus fringe benefits and a 33% allowance for general overhead items, taxes, insurance, supervision, and small tools.

All labor hours for installation were estimated using the *Labor Estimating Manual,* published by Mechanical Contractors Association of America, Inc., where covered by that book. Where labor for any task was not covered by that book, we used a variety of other manuals and books (see References).

## Labor Correction Factors

Different authorities have different methods of adjusting production for installing material and equipment at elevated heights and in high-rise buildings. While this book does not show breakdowns of material and labor costs, the following rules of thumb may be helpful to the estimator:

Floor to installation height:
Add 1.5% to the labor cost for each foot of height above 12 ft.

High-rise construction:

Add 1.0% to the labor cost for each floor for building floors 5 through 20.

Add 3% to the labor cost for each floor above 20.

## Building Construction Cost Escalation

As a guide to past increases in costs and possible future escalation, we have included Figure 2.1, Colorado Building Construction Cost Increases Averages, and Table 2.1, Typical Nonresidential Building Cost Index Summary, both published by the Hensel Phelps Construction Company, Greeley, Colorado.

## Building Cost Adjustment Indices

Researching several information sources shows some difference of opinion as to the adjustment index for various cities. While Dodge shows Denver's combined labor and material adjustment index as 1.02, one source shows the mechanical trades index for Denver as .96. The comparison of other cities' index numbers also shows some differences. In any case, judgment must be applied, and if either labor or material in your area varies drastically from the average, the values in this manual should be carefully scrutinized.

Included for the estimators' use are the adjustment indices from the 1986 *Dodge Manual for Building Construction Pricing and Scheduling* (Table 2.2).

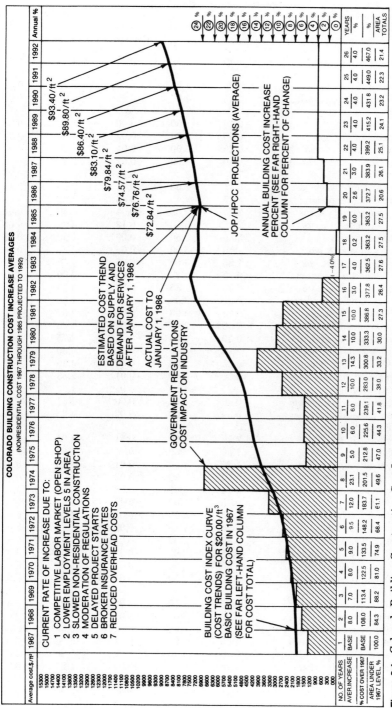

**Figure 2.1** Colorado Building Construction Cost Increases Averages.

TABLE 2.1   Typical Nonresidential Building Cost Index Summary

| Base category of work item description | Base cost | Average, 12 mo. | 1985 increase | Average, 12 mo. | 1986 increase |
|---|---|---|---|---|---|
| Excavations and grading | 1.50 | -2.12 | -0.03 | 0.00 | 0.00 |
| Drilled pier foundation | 3.00 | -1.50 | -0.05 | 0.75 | 0.02 |
| Asphalt paving and base | 0.25 | 1.75 | 0.00 | 3.25 | 0.01 |
| Chain link fencing work | 0.25 | 2.67 | 0.01 | 4.00 | 0.01 |
| Building wrecking/ demolition | 0.25 | -1.00 | 0.00 | 1.00 | 0.00 |
| Landscaping and sodding | 0.50 | 3.33 | 0.02 | 1.67 | 0.01 |
| Lawn irrigation systems | 0.25 | 5.00 | 0.01 | 4.00 | 0.01 |
| Underground and utilities | 0.50 | 3.00 | 0.02 | 3.50 | 0.02 |
| Concrete materials/work | 7.25 | -2.00 | -0.15 | 3.76 | 0.27 |
| Forms and formwork cost | 6.75 | -2.00 | -0.14 | 2.00 | 0.14 |
| Reinforcing and placing | 4.75 | -7.50 | -0.36 | 0.67 | 0.03 |
| Post-tensioning systems | 2.00 | -2.00 | -0.04 | -0.50 | -0.01 |
| Architect. precast costs | 2.25 | -5.50 | -0.12 | -2.00 | -0.05 |
| Structural prestressed | 3.50 | -2.50 | -0.09 | 1.50 | 0.05 |
| Masonry facebrick work | 1.50 | -3.00 | -0.05 | 2.67 | 0.04 |
| Masonry blockwork costs | 1.25 | -3.60 | -0.05 | 3.50 | 0.04 |
| Granite masonry systems | 0.25 | 4.33 | 0.01 | 4.00 | 0.01 |
| Structural steel system | 3.50 | 1.35 | 0.05 | 3.50 | 0.12 |
| Metal decking and sidings | 0.25 | 2.00 | 0.01 | 1.00 | 0.00 |
| Steel bar joist systems | 0.25 | 4.00 | 0.01 | 0.67 | 0.00 |
| Miscellaneous metalwork | 0.50 | -2.00 | -0.01 | 5.00 | 0.03 |
| Rough carpentry systems | 0.50 | -1.25 | -0.01 | 3.75 | 0.02 |
| Millwork and door costs | 0.75 | 1.00 | 0.01 | 3.00 | 0.02 |
| Caulking/sealant system | 0.25 | 3.00 | 0.01 | 4.50 | 0.01 |
| Roofing and Insulations | 0.50 | -1.67 | -0.01 | 2.83 | 0.01 |
| Outside bldg. sheet metals | 0.25 | 4.00 | 0.01 | 1.00 | 0.00 |
| Aluminum skylite system | 0.25 | -1.67 | 0.00 | 1.67 | 0.00 |
| Hollow metal door/frame | 0.50 | 4.00 | 0.02 | 4.33 | 0.02 |
| Glazing and aluminum work | 2.75 | 4.00 | 0.11 | 4.00 | 0.11 |
| Special curtain walls | 1.00 | 3.50 | 0.04 | 1.67 | 0.02 |
| Acoustical ceiling work | 1.25 | 1.10 | 0.01 | 3.23 | 0.04 |
| Lathing/plastering work | 1.00 | 2.00 | 0.02 | 4.50 | 0.05 |
| Gypsum drywall systems | 2.75 | 3.00 | 0.08 | 3.06 | 0.08 |
| Ceramic/quarry tilework | 0.75 | 5.00 | 0.04 | 7.50 | 0.06 |
| Resilient flooring work | 0.25 | -5.00 | -0.01 | 2.50 | 0.01 |
| Painting and vinyl work | 1.75 | 1.46 | 0.03 | 1.72 | 0.03 |
| Terrazzo flooring costs | 0.25 | 2.00 | 0.01 | 1.00 | 0.00 |
| Finish hardware systems | 0.75 | 3.67 | 0.03 | 3.67 | 0.03 |
| Movable metal partition | 0.75 | 5.00 | 0.04 | 5.00 | 0.04 |
| School equipment system | 1.50 | 2.50 | 0.04 | 5.00 | 0.08 |
| Food service equipment cost | 0.25 | 1.50 | 0.00 | 0.00 | 0.00 |
| Security/jail equipment | 0.25 | 4.00 | 0.01 | 3.00 | 0.01 |
| Laboratory casework/top | 0.50 | 3.00 | 0.02 | 3.00 | 0.02 |
| Carpeting and draperies | 1.00 | 1.00 | 0.01 | 5.50 | 0.06 |
| Elevators and escalator | 2.25 | 1.40 | 0.03 | 3.20 | 0.07 |
| Plumbing system/fixture | 8.25 | -0.60 | -0.05 | 1.80 | 0.15 |
| Heat/vent/air cond. work | 14.00 | 0.33 | 0.05 | 3.75 | 0.53 |
| Fire protection systems | 1.75 | 1.30 | 0.02 | 3.00 | 0.05 |
| Electrical systems work | 10.50 | 3.25 | 0.34 | 2.33 | 0.24 |
| General conditions work | 3.00 | 1.50 | 0.05 | 3.00 | 0.09 |
| J.G.P. adjustments made | | | 0.02 | | 0.00 |
| | 100.00 | | 0.00 | | 2.60 |

Prepared January 1, 1986 by Jerry Morgensen and Jerry Pope.

**TABLE 2.2. Adjustment Indices**

| Location | | Labor | Material | General | Location | | Labor | Material | General |
|---|---|---|---|---|---|---|---|---|---|
| San Juan | PR | 0.29 | 1.01 | 0.65 | Rutland | VT | 0.65 | 1.01 | 0.83 |
| Springfield | MA | 0.91 | 0.96 | 0.93 | Saint Johnsbury | VT | 0.52 | 1.01 | 0.76 |
| Pittsfield | MA | 0.81 | 0.95 | 0.88 | Hartford | CT | 0.85 | 1.02 | 0.93 |
| Fitchburg | MA | 0.95 | 1.01 | 0.98 | New London | CT | 0.83 | 0.99 | 0.91 |
| Rutland | MA | 0.97 | 1.00 | 0.98 | Norwich | CT | 0.81 | 0.99 | 0.90 |
| Shrewsbury | MA | 0.97 | 1.00 | 0.98 | New Haven | CT | 0.99 | 1.05 | 1.02 |
| Webster | MA | 0.97 | 1.00 | 0.98 | Bridgeport | CT | 0.96 | 0.99 | 0.97 |
| Worcester | MA | 0.97 | 1.00 | 0.98 | Waterbury | CT | 0.93 | 1.02 | 0.97 |
| Concord | MA | 1.08 | 1.11 | 1.09 | Torrington | CT | 0.90 | 0.99 | 0.94 |
| Lowell | MA | 0.95 | 1.06 | 1.00 | Greenwich | CT | 1.09 | 1.08 | 1.08 |
| Lynn | MA | 1.08 | 1.11 | 1.09 | Stamford | CT | 1.09 | 1.08 | 1.08 |
| Salem | MA | 0.92 | 1.01 | 0.96 | Orange | NJ | 0.96 | 0.95 | 0.95 |
| Norwood | MA | 1.08 | 1.11 | 1.09 | Newark | NJ | 1.05 | 0.96 | 1.00 |
| Boston | MA | 1.08 | 1.11 | 1.09 | Jersey City | NJ | 0.98 | 0.98 | 0.98 |
| Waltham | MA | 0.96 | 1.09 | 1.02 | Paterson | NJ | 0.98 | 0.93 | 0.95 |
| Lexington | MA | 1.08 | 1.11 | 1.09 | Hackensack | NJ | 0.95 | 0.96 | 0.95 |
| Milton | MA | 1.08 | 1.11 | 1.09 | Asbury Park | NJ | 0.95 | 0.91 | 0.93 |
| Brockton | MA | 0.99 | 1.00 | 0.99 | Morristown | NJ | 0.97 | 0.96 | 0.96 |
| Hyannis | MA | 0.92 | 1.06 | 0.99 | Cherry Hill | NJ | 0.97 | 0.91 | 0.94 |
| New Bedford | MA | 0.98 | 1.01 | 0.99 | Camden | NJ | 0.97 | 0.91 | 0.94 |
| Providence | RI | 0.82 | 0.98 | 0.90 | Wildwood | NJ | 1.02 | 1.06 | 1.04 |
| Manchester | NH | 0.80 | 1.00 | 0.90 | Bridgeton | NJ | 1.04 | 1.06 | 1.05 |
| Claremont | NH | 0.71 | 1.05 | 0.88 | Atlantic City | NJ | 1.08 | 1.06 | 1.07 |
| Portsmouth | NH | 0.72 | 1.02 | 0.87 | Trenton | NJ | 0.95 | 0.93 | 0.94 |
| Portland | ME | 0.72 | 1.05 | 0.88 | Toms River | NJ | 0.97 | 1.00 | 0.98 |
| Augusta | ME | 0.73 | 0.96 | 0.84 | East Brunswick | NJ | 0.97 | 0.99 | 0.98 |
| Bangor | ME | 0.68 | 0.86 | 0.77 | Flemington | NJ | 0.95 | 0.97 | 0.96 |
| Machias | ME | 0.57 | 1.17 | 0.87 | Phillipsburg | NJ | 0.90 | 1.05 | 0.97 |
| Presque Isle | ME | 0.50 | 1.18 | 0.84 | New Brunswick | NJ | 0.96 | 0.97 | 0.96 |
| Burlington | VT | 0.79 | 1.03 | 0.91 | New York City | NY | 1.26 | 1.12 | 1.19 |
| Montpelier | VT | 0.80 | 1.03 | 0.91 | Staten Island | NY | 1.26 | 1.12 | 1.19 |

**TABLE 2.2. Adjustment Indices** (Continued)

| Location | | Labor | Material | General | Location | | Labor | Material | General |
|---|---|---|---|---|---|---|---|---|---|
| Bronx | NY | 1.26 | 1.12 | 1.19 | Dubois | PA | 0.94 | 0.89 | 0.91 |
| Mamaroneck | NY | 1.07 | 1.19 | 1.13 | Johnstown | PA | 0.96 | 0.95 | 0.95 |
| Rye | NY | 1.07 | 1.19 | 1.13 | Newcastle | PA | 1.00 | 0.97 | 0.98 |
| White Plains | NY | 1.07 | 1.19 | 1.13 | Ellwood City | PA | 1.00 | 0.99 | 0.99 |
| Brooklyn | NY | 1.26 | 1.12 | 1.19 | Dayton | PA | 0.96 | 0.95 | 0.95 |
| Queens | NY | 1.26 | 1.12 | 1.19 | Sagamore | PA | 0.96 | 0.95 | 0.95 |
| Nassau-Suffolk | NY | 0.96 | 1.01 | 0.98 | Waterford | PA | 0.89 | 0.94 | 0.91 |
| Albany | NY | 0.84 | 0.91 | 0.87 | Erie | PA | 0.89 | 0.94 | 0.91 |
| Schenectady | NY | 0.85 | 0.89 | 0.87 | Altoona | PA | 0.86 | 0.81 | 0.83 |
| Kingston | NY | 0.92 | 0.98 | 0.95 | Camp Hill | PA | 0.82 | 0.90 | 0.88 |
| Poughkeepsie | NY | 0.94 | 0.94 | 0.94 | Harrisburg | PA | 0.82 | 0.90 | 0.86 |
| Glens Falls | NY | 0.79 | 0.94 | 0.86 | Bart | PA | 0.77 | 0.84 | 0.80 |
| Plattsburgh | NY | 0.78 | 0.96 | 0.87 | Kirkwood | PA | 0.77 | 0.84 | 0.80 |
| Syracuse | NY | 0.89 | 1.07 | 0.98 | New Holland | PA | 0.77 | 0.84 | 0.80 |
| Utica | NY | 0.86 | 0.98 | 0.92 | Quaryville | PA | 0.77 | 0.84 | 0.80 |
| Watertown | NY | 0.88 | 0.96 | 0.92 | Lancaster | PA | 0.77 | 0.84 | 0.80 |
| Potsdam | NY | 0.87 | 0.95 | 0.91 | Williamsport | PA | 0.82 | 0.91 | 0.86 |
| Binghamton | NY | 0.78 | 0.94 | 0.86 | Allentown | PA | 0.82 | 1.07 | 0.94 |
| Buffalo | NY | 1.01 | 0.86 | 0.93 | Scranton | PA | 0.86 | 0.93 | 0.89 |
| Rochester | NY | 0.93 | 0.95 | 0.94 | Wilkes-Barre | PA | 0.86 | 0.91 | 0.88 |
| Jamestown | NY | 0.87 | 0.97 | 0.92 | Levittown | PA | 1.02 | 1.17 | 1.09 |
| Elmira | NY | 0.79 | 0.96 | 0.87 | Philadelphia | PA | 1.02 | 1.17 | 1.09 |
| Pittsburgh | PA | 1.00 | 0.99 | 0.99 | Coatsville | PA | 1.02 | 1.17 | 1.09 |
| Washington | PA | 0.97 | 0.89 | 0.93 | Oxford | PA | 0.77 | 0.84 | 0.80 |
| Ursina | PA | 0.96 | 0.95 | 0.95 | Westchester | PA | 1.02 | 1.17 | 1.09 |
| Somerset | PA | 0.96 | 0.95 | 0.95 | King of Prussia | PA | 1.02 | 1.17 | 1.09 |
| Jennerstown | PA | 0.96 | 0.95 | 0.95 | Pottstown | PA | 1.02 | 1.17 | 1.09 |
| Greenburg | PA | 1.00 | 0.99 | 0.99 | Adamstown | PA | 0.82 | 1.09 | 0.95 |
| Murrinsville | PA | 1.00 | 0.97 | 0.98 | Reading | PA | 0.82 | 1.09 | 0.95 |
| Vandergrift | PA | 1.00 | 0.99 | 0.99 | Middletown | PA | 0.87 | 1.00 | 0.93 |
| Indiana | PA | 0.96 | 0.95 | 0.95 | Wilmington | DE | 0.87 | 1.00 | 0.93 |

| City | State | | | |
|---|---|---|---|---|
| Dover | DE | 0.90 | 0.96 | 0.93 |
| Washington | DC | 0.84 | 0.92 | 0.88 |
| Silver Springs | MD | 0.85 | 1.00 | 0.92 |
| Edgewood | MD | 0.87 | 1.00 | 0.93 |
| Glen Burnie | MD | 0.87 | 1.00 | 0.93 |
| Randallstown | MD | 0.87 | 1.00 | 0.93 |
| Reisterstown | MD | 0.87 | 1.00 | 0.93 |
| Baltimore | MD | 0.87 | 0.95 | 0.90 |
| Annapolis | MD | 0.85 | 1.00 | 0.91 |
| Cumberland | MD | 0.82 | 0.95 | 0.82 |
| Easton | MD | 0.70 | 0.95 | 0.84 |
| Cambridge | MD | 0.74 | 0.94 | 0.87 |
| Hagerstown | MD | 0.80 | 0.97 | 0.86 |
| Salisbury | MD | 0.75 | 0.99 | 0.87 |
| Warrenton | VA | 0.76 | 1.00 | 0.88 |
| Winchester | VA | 0.77 | 1.10 | 0.99 |
| Charlottesville | VA | 0.88 | 1.01 | 0.83 |
| Richmond | VA | 0.65 | 0.97 | 0.82 |
| Suffolk | VA | 0.68 | 0.96 | 0.81 |
| Norfolk | VA | 0.66 | 0.94 | 0.76 |
| Roanoke | VA | 0.58 | 0.94 | 0.76 |
| Lynchburg | VA | 0.58 | 0.94 | 0.65 |
| Danville | VA | 0.37 | 0.95 | 0.83 |
| Bluefield | WV | 0.72 | 1.07 | 1.00 |
| Charleston | WV | 0.93 | 1.13 | 1.04 |
| Huntington | WV | 0.96 | 0.97 | 0.93 |
| Beckley | WV | 0.90 | 0.98 | 0.91 |
| Parkersburg | WV | 0.84 | 0.95 | 0.89 |
| Clarksburg | WV | 0.84 | 0.96 | 0.90 |
| Fairmont | WV | 0.84 | 0.95 | 0.77 |
| Winston-Salem | NC | 0.59 | 0.94 | 0.72 |
| Greensboro | NC | 0.51 | 0.85 | 0.63 |
| Goldsboro | NC | 0.42 | 0.96 | 0.71 |
| Raleigh | NC | 0.47 | | |
| Rocky Mount | NC | 0.58 | 0.87 | 0.72 |
| Charlotte | NC | 0.61 | 0.98 | 0.79 |
| Fayetteville | NC | 0.51 | 0.88 | 0.69 |
| Wilmington | NC | 0.51 | 0.98 | 0.74 |
| New Bern | NC | 0.47 | 0.89 | 0.68 |
| Lenoir | NC | 0.34 | 0.92 | 0.63 |
| Asheville | NC | 0.37 | 0.89 | 0.63 |
| Orangeburg | SC | 0.51 | 0.96 | 0.73 |
| Columbia | SC | 0.55 | 0.91 | 0.73 |
| Spartanburg | SC | 0.35 | 0.95 | 0.65 |
| Charleston | SC | 0.55 | 0.92 | 0.73 |
| Florence | SC | 0.58 | 0.91 | 0.74 |
| Myrtle Beach | SC | 0.42 | 0.98 | 0.70 |
| Greenville | SC | 0.49 | 1.05 | 0.77 |
| Anderson | SC | 0.49 | 1.02 | 0.75 |
| Greenwood | SC | 0.49 | 1.02 | 0.75 |
| Beaufort | SC | 0.54 | 1.05 | 0.79 |
| Rome | GA | 0.72 | 0.86 | 0.79 |
| Covington | GA | 0.76 | 0.91 | 0.83 |
| Griffin | GA | 0.76 | 0.91 | 0.83 |
| Hogansville | GA | 0.65 | 0.89 | 0.77 |
| Jackson | GA | 0.76 | 0.91 | 0.83 |
| Milner | GA | 0.53 | 0.89 | 0.71 |
| Newborn | GA | 0.76 | 0.91 | 0.83 |
| Newnan | GA | 0.76 | 0.91 | 0.83 |
| Zebulon | GA | 0.76 | 0.91 | 0.83 |
| Atlanta | GA | 0.74 | 0.91 | 0.82 |
| Gainesville | GA | 0.44 | 0.90 | 0.67 |
| Athens | GA | 0.39 | 0.85 | 0.62 |
| Calhoun | GA | 0.72 | 0.86 | 0.79 |
| Carters | GA | 0.72 | 0.86 | 0.79 |
| Augusta | GA | 0.61 | 0.97 | 0.79 |
| Dublin | GA | 0.54 | 0.93 | 0.73 |
| Macon | GA | 0.53 | 0.89 | 0.71 |

TABLE 2.2. Adjustment Indices (Continued)

| Location | | Labor | Material | General | Location | | Labor | Material | General |
|---|---|---|---|---|---|---|---|---|---|
| Savannah | GA | 0.64 | 1.05 | 0.84 | Montgomery | AL | 0.65 | 0.81 | 0.73 |
| Waycross | GA | 0.34 | 0.94 | 0.64 | Anniston | AL | 0.72 | 0.83 | 0.77 |
| Brunswick | GA | 0.68 | 0.95 | 0.81 | Ashland | AL | 0.72 | 0.83 | 0.77 |
| Valdosta | GA | 0.33 | 0.89 | 0.61 | Bynum | AL | 0.72 | 0.83 | 0.77 |
| Albany | GA | 0.61 | 0.92 | 0.76 | Munford | AL | 0.72 | 0.83 | 0.77 |
| Columbus | GA | 0.45 | 0.89 | 0.67 | Dothan | AL | 0.51 | 1.11 | 0.81 |
| Daytona Beach | FL | 0.74 | 0.88 | 0.81 | Mobile | AL | 0.76 | 0.90 | 0.83 |
| Jacksonville | FL | 0.69 | 0.87 | 0.78 | Selma | AL | 0.58 | 0.91 | 0.74 |
| Tallahassee | FL | 0.62 | 0.87 | 0.74 | Clarksville | TN | 0.70 | 0.85 | 0.77 |
| Panama City | FL | 0.57 | 0.85 | 0.71 | Nashville | TN | 0.66 | 0.78 | 0.72 |
| Pensacola | FL | 0.69 | 0.85 | 0.77 | Chattanooga | TN | 0.70 | 0.82 | 0.76 |
| Gainesville | FL | 0.67 | 0.89 | 0.78 | Johnson City | TN | 0.61 | 0.92 | 0.76 |
| Orlando | FL | 0.75 | 0.87 | 0.81 | Kingsport | TN | 0.61 | 0.92 | 0.76 |
| Cocoa | FL | 0.74 | 0.86 | 0.80 | Knoxville | TN | 0.65 | 0.83 | 0.74 |
| Miami | FL | 0.81 | 0.87 | 0.84 | Memphis | TN | 0.79 | 0.89 | 0.84 |
| Fort Lauderdale | FL | 0.81 | 0.87 | 0.84 | Union City | TN | 0.57 | 0.91 | 0.74 |
| West Palm Beach | FL | 0.80 | 0.88 | 0.84 | Jackson | TN | 0.55 | 0.88 | 0.71 |
| Fort Pierce | FL | 0.80 | 0.87 | 0.83 | Columbia | TN | 0.46 | 0.92 | 0.69 |
| Sarasota | FL | 0.75 | 0.88 | 0.81 | Cookeville | TN | 0.50 | 0.84 | 0.67 |
| Tampa | FL | 0.76 | 0.88 | 0.82 | Clarksdale | MS | 0.53 | 1.02 | 0.77 |
| Saint Petersburg | FL | 0.72 | 0.88 | 0.80 | Greenville | MS | 0.53 | 0.93 | 0.73 |
| Lakeland | FL | 0.73 | 0.88 | 0.80 | Tupelo | MS | 0.57 | 0.95 | 0.76 |
| Fort Myers | FL | 0.70 | 0.88 | 0.79 | Greenwood | MS | 0.38 | 0.99 | 0.68 |
| Naples | FL | 0.75 | 0.87 | 0.81 | Natchez | MS | 0.46 | 0.89 | 0.67 |
| Brent | AL | 0.68 | 0.84 | 0.76 | Jackson | MS | 0.67 | 0.91 | 0.79 |
| Birmingham | AL | 0.72 | 0.83 | 0.77 | Meridian | MS | 0.44 | 0.94 | 0.69 |
| Tuscaloosa | AL | 0.68 | 0.84 | 0.76 | Hattiesburg | MS | 0.65 | 1.07 | 0.86 |
| Jasper | AL | 0.72 | 0.83 | 0.77 | Gulfport | MS | 0.63 | 0.98 | 0.80 |
| Florence | AL | 0.67 | 0.93 | 0.80 | Columbus | MS | 0.47 | 0.88 | 0.67 |
| Huntsville | AL | 0.70 | 0.81 | 0.75 | Louisville | KY | 0.78 | 0.86 | 0.82 |
| Gadsden | AL | 0.65 | 0.86 | 0.75 | Lexington | KY | 0.83 | 0.92 | 0.87 |

| City | State | | | |
|------|-------|------|------|------|
| Middlesboro | KY | 0.56 | 0.88 | 0.72 |
| Covington | KY | 0.98 | 0.93 | 0.95 |
| Ashland | KY | 0.84 | 0.87 | 0.85 |
| Pikesville | KY | 0.66 | 1.07 | 0.86 |
| Paducah | KY | 0.71 | 0.94 | 0.82 |
| Bowling Green | KY | 0.53 | 0.84 | 0.68 |
| Owensboro | KY | 0.77 | 1.00 | 0.88 |
| Madisonville | KY | 0.72 | 0.82 | 0.77 |
| Somerset | KY | 0.47 | 0.84 | 0.65 |
| Elizabethtown | KY | 0.54 | 0.87 | 0.70 |
| Columbus | OH | 0.94 | 1.02 | 0.98 |
| Toledo | OH | 1.00 | 0.92 | 0.96 |
| Zanesville | OH | 0.86 | 0.92 | 0.89 |
| Steubenville | OH | 0.82 | 0.96 | 0.89 |
| Elyria | OH | 1.09 | 1.00 | 1.04 |
| North Olmsted | OH | 1.09 | 1.00 | 1.04 |
| Painesville | OH | 1.03 | 0.98 | 1.00 |
| Twinsburg | OH | 1.09 | 1.00 | 1.04 |
| Cleveland | OH | 1.09 | 1.00 | 1.04 |
| Richfield | OH | 1.09 | 1.00 | 1.04 |
| Akron | OH | 0.96 | 1.00 | 0.98 |
| Cortland | OH | 0.96 | 0.96 | 0.96 |
| Mineral Ridge | OH | 0.96 | 0.96 | 0.96 |
| Niles | OH | 0.96 | 0.96 | 0.96 |
| Warren | OH | 0.96 | 0.96 | 0.96 |
| Youngstown | OH | 0.96 | 0.96 | 0.96 |
| Canton | OH | 0.90 | 1.00 | 0.95 |
| Polk | OH | 0.98 | 0.94 | 0.96 |
| Sandusky | OH | 0.98 | 0.99 | 0.98 |
| Mansfield | OH | 0.98 | 0.94 | 0.96 |
| Decatur | OH | 0.90 | 0.92 | 0.91 |
| Cincinnati | OH | 0.99 | 0.96 | 0.97 |
| Brookville | OH | 0.94 | 0.92 | 0.93 |
| Eaton | OH | 0.92 | 0.86 | 0.89 |

| City | State | | | |
|------|-------|------|------|------|
| Germantown | OH | 0.94 | 0.92 | 0.93 |
| Lewisburg | OH | 0.99 | 0.96 | 0.97 |
| New Paris | OH | 0.99 | 0.96 | 0.97 |
| SpringValley | OH | 0.94 | 0.92 | 0.93 |
| Xenia | OH | 0.94 | 0.92 | 0.93 |
| Dayton | OH | 0.94 | 0.92 | 0.93 |
| Portsmouth | OH | 0.90 | 0.92 | 0.91 |
| Athens | OH | 0.74 | 0.92 | 0.83 |
| Lima | OH | 0.90 | 0.95 | 0.92 |
| Indianapolis | IN | 0.93 | 1.04 | 0.98 |
| Gary | IN | 0.93 | 1.01 | 0.97 |
| South Bend | IN | 0.91 | 0.87 | 0.89 |
| Fort Wayne | IN | 0.81 | 0.99 | 0.90 |
| Kokomo | IN | 0.79 | 1.00 | 0.89 |
| New Albany | IN | 0.81 | 0.85 | 0.83 |
| Greensburg | IN | 0.93 | 0.90 | 0.91 |
| Muncie | IN | 0.84 | 0.94 | 0.89 |
| Richmond | IN | 0.86 | 1.01 | 0.93 |
| Bloomington | IN | 0.83 | 0.89 | 0.86 |
| Evansville | IN | 0.92 | 1.02 | 0.97 |
| Terre Haute | IN | 0.78 | 1.03 | 0.90 |
| Lafayette | IN | 0.80 | 0.90 | 0.85 |
| Port Huron | MI | 0.94 | 1.01 | 0.97 |
| Ann Arbor | MI | 1.06 | 1.13 | 1.09 |
| Hamburg | MI | 1.06 | 1.13 | 1.09 |
| Luna Pier | MI | 1.06 | 1.13 | 1.09 |
| Samaria | MI | 1.06 | 1.13 | 1.09 |
| Detroit | MI | 1.06 | 1.13 | 1.09 |
| Fenton | MI | 0.99 | 1.08 | 1.03 |
| Goodrich | MI | 0.99 | 1.08 | 1.03 |
| Hadley | MI | 1.99 | 1.08 | 1.03 |
| Imlay City | MI | 0.99 | 1.08 | 1.03 |
| Flint | MI | 0.99 | 1.08 | 1.03 |
| Charlotte | MI | 0.86 | 1.01 | 0.93 |

TABLE 2.2. Adjustment Indices (Continued)

| Location | | Labor | Material | General |
|---|---|---|---|---|
| Portland | MI | 0.86 | 1.01 | 0.93 |
| St. Johns | MI | 0.86 | 1.01 | 0.93 |
| Lansing | MI | 0.86 | 1.01 | 0.93 |
| Kalamazoo | MI | 0.81 | 1.01 | 0.91 |
| Adrian | MI | 1.06 | 1.13 | 1.09 |
| Clinton | MI | 1.06 | 1.13 | 1.09 |
| Petersburg | MI | 1.06 | 1.13 | 1.09 |
| Stockbridge | MI | 0.86 | 1.01 | 0.93 |
| Muskegon | MI | 0.79 | 1.01 | 0.90 |
| Grand Rapids | MI | 0.79 | 1.02 | 0.90 |
| Traverse City | MI | 0.75 | 1.04 | 0.89 |
| Alpena | MI | 0.78 | 1.00 | 0.89 |
| Petosky | MI | 0.75 | 1.01 | 0.88 |
| Sault Sainte Marie | MI | 0.76 | 0.99 | 0.87 |
| Marquette | MI | 0.78 | 0.94 | 0.86 |
| Ironwood | MI | 0.69 | 0.99 | 0.84 |
| Marshalltown | IA | 0.75 | 0.96 | 0.85 |
| Des Moines | IA | 0.80 | 0.95 | 0.87 |
| Mason City | IA | 0.63 | 0.93 | 0.78 |
| Fort Dodge | IA | 0.74 | 1.09 | 0.91 |
| Waterloo | IA | 0.70 | 1.07 | 0.88 |
| Creston | IA | 0.63 | 1.08 | 0.85 |
| Sioux City | IA | 0.73 | 0.93 | 0.83 |
| Spencer | IA | 0.59 | 0.93 | 0.76 |
| Council Bluffs | IA | 0.77 | 0.84 | 0.80 |
| Dubuque | IA | 0.74 | 0.97 | 0.85 |
| Cedar Rapids | IA | 0.80 | 0.97 | 0.88 |
| Ottumwa | IA | 0.78 | 0.96 | 0.87 |
| Burlington | IA | 0.82 | 0.98 | 0.90 |
| Davenport | IA | 0.87 | 0.95 | 0.91 |
| Sheboygan | WI | 0.80 | 0.92 | 0.86 |

| Location | | Labor | Material | General |
|---|---|---|---|---|
| Milwaukee | WI | 0.93 | 1.05 | 0.99 |
| Janesville | WI | 0.88 | 0.90 | 0.89 |
| Madison | WI | 0.83 | 0.92 | 0.87 |
| Marinette | WI | 0.85 | 0.92 | 0.88 |
| Green Bay | WI | 0.82 | 0.95 | 0.88 |
| Wausau | WI | 0.76 | 0.88 | 0.82 |
| Rhinelander | WI | 0.86 | 0.99 | 0.92 |
| La Crosse | WI | 0.81 | 1.08 | 0.94 |
| Eau Claire | WI | 0.81 | 0.98 | 0.89 |
| Rice Lake | WI | 0.83 | 0.96 | 0.89 |
| Fond du Lac | WI | 0.83 | 1.00 | 0.91 |
| Northfield | MN | 0.94 | 1.14 | 1.04 |
| Owatouna | MN | 0.89 | 0.97 | 0.93 |
| St. Paul | MN | 0.94 | 1.14 | 1.04 |
| Chanhassen | MN | 0.94 | 1.14 | 1.04 |
| Hutchinson | MN | 0.94 | 1.14 | 1.04 |
| Kimball Prairie | MN | 0.83 | 0.98 | 0.90 |
| Princeton | MN | 0.83 | 0.98 | 0.90 |
| Winthrop | MN | 0.78 | 0.93 | 0.85 |
| Minneapolis | MN | 0.94 | 1.13 | 1.03 |
| Virginia | MN | 0.87 | 1.11 | 0.99 |
| Duluth | MN | 0.87 | 0.97 | 0.92 |
| Rochester | MN | 0.89 | 0.97 | 0.93 |
| Winona | MN | 0.89 | 0.97 | 0.93 |
| Mankato | MN | 0.78 | 0.93 | 0.85 |
| Montevideo | MN | 0.60 | 0.95 | 0.77 |
| Saint Cloud | MN | 0.83 | 0.98 | 0.90 |
| Brainerd | MN | 0.80 | 0.94 | 0.87 |
| Yankton | SD | 0.60 | 0.99 | 0.79 |
| Sioux Falls | SD | 0.65 | 0.97 | 0.81 |
| Watertown | SD | 0.62 | 0.97 | 0.79 |

| City | State | | | |
|---|---|---|---|---|
| Chamberlain | SD | 0.66 | 1.00 | 0.83 |
| Huron | SD | 0.66 | 1.01 | 0.83 |
| Aberdeen | SD | 0.48 | 0.91 | 0.69 |
| Pierre | SD | 0.43 | 1.17 | 0.80 |
| Rapid City | SD | 0.59 | 0.94 | 0.76 |
| Fargo | ND | 0.67 | 0.96 | 0.81 |
| Grand Forks | ND | 0.60 | 0.99 | 0.79 |
| Devils Lake | ND | 0.53 | 1.08 | 0.80 |
| Jamestown | ND | 0.63 | 0.92 | 0.77 |
| Bismarck | ND | 0.69 | 1.07 | 0.88 |
| Minot | ND | 0.70 | 0.98 | 0.84 |
| Williston | ND | 0.49 | 1.04 | 0.76 |
| Billings | MT | 0.84 | 1.00 | 0.92 |
| Texarkana | TX | 0.67 | 0.93 | 0.80 |
| Tyler | TX | 0.60 | 0.92 | 0.76 |
| Nacogdoches | TX | 0.49 | 0.91 | 0.70 |
| Bridgeport | TX | 0.82 | 0.98 | 0.90 |
| Cleburne | TX | 0.82 | 0.98 | 0.90 |
| White Settlement | TX | 0.82 | 0.98 | 0.90 |
| Gainesville | TX | 0.60 | 0.87 | 0.73 |
| Wichita Falls | TX | 0.71 | 0.97 | 0.84 |
| Dawson | TX | 0.64 | 0.88 | 0.76 |
| Hillsboro | TX | 0.64 | 0.88 | 0.76 |
| Whitney | TX | 0.64 | 0.88 | 0.76 |
| Waco | TX | 0.64 | 0.89 | 0.76 |
| San Angelo | TX | 0.56 | 0.89 | 0.72 |
| Houston | TX | 0.89 | 0.84 | 0.86 |
| Granbury | TX | 0.82 | 0.98 | 0.90 |
| Beaumont | TX | 0.85 | 0.86 | 0.85 |
| San Antonio | TX | 0.72 | 0.90 | 0.81 |
| Corpus Christi | TX | 0.67 | 0.91 | 0.79 |
| Brownsville | TX | 0.48 | 0.95 | 0.71 |
| Harlingen | TX | 0.49 | 0.94 | 0.71 |
| Austin | TX | 0.68 | 0.98 | 0.83 |
| Del Rio | TX | 0.35 | 0.94 | 0.64 |
| Pampa | TX | 0.66 | 1.00 | 0.83 |
| Amarillo | TX | 0.71 | 1.02 | 0.86 |
| Lubbock | TX | 0.58 | 0.99 | 0.78 |
| Abilene | TX | 0.57 | 1.06 | 0.81 |
| Odessa | TX | 0.60 | 0.91 | 0.75 |
| El Paso | TX | 0.55 | 0.97 | 0.76 |
| Castle Rock | CO | 0.96 | 1.09 | 1.02 |
| Strasburg | CO | 0.96 | 1.09 | 1.02 |
| Denver | CO | 0.96 | 1.09 | 1.02 |
| Green Mountain | CO | 0.93 | 1.18 | 1.05 |
| Boulder | CO | 0.96 | 1.09 | 1.02 |
| Central City | CO | 0.96 | 1.09 | 1.02 |
| Allens Park | CO | 0.90 | 1.01 | 0.95 |
| Fort Collins | CO | 0.96 | 1.09 | 1.02 |
| Greeley | CO | 0.91 | 1.07 | 0.99 |
| Sterling | CO | 0.93 | 1.18 | 1.05 |
| Calhan | CO | 0.93 | 1.18 | 1.05 |
| Florissant | CO | 0.93 | 1.18 | 1.05 |
| Lake George | CO | 0.93 | 1.18 | 1.05 |
| Ramah | CO | 0.93 | 1.18 | 1.05 |
| Colorado Springs | CO | 0.80 | 1.10 | 0.95 |
| Pueblo | CO | 0.74 | 0.97 | 0.85 |
| La Junta | CO | 0.70 | 1.06 | 0.88 |
| Trinidad | CO | 0.93 | 0.93 | 0.93 |
| Montrose | CO | 0.93 | 0.96 | 0.94 |
| Grand Junction | CO | 0.82 | 1.15 | 0.98 |
| Cheyenne | WY | 0.72 | 1.23 | 0.97 |
| Thermopolis | WY | 0.81 | 1.01 | 0.91 |
| Casper | WY | 0.89 | 1.07 | 0.98 |
| Gillette | WY | 0.85 | 1.05 | 0.95 |
| Sheridan | WY | 0.83 | 1.10 | 0.96 |
| Rock Springs | WY | 0.84 | 0.94 | 0.89 |
| Pocatello | ID | | | |

**TABLE 2.2. Adjustment Indices (Continued)**

| Location | | Labor | Material | General | Location | | Labor | Material | General |
|---|---|---|---|---|---|---|---|---|---|
| Twin Falls | ID | 0.87 | 0.99 | 0.93 | Canoga Park | CA | 1.28 | 1.01 | 1.14 |
| Idaho Falls | ID | 0.84 | 0.95 | 0.89 | Encino | CA | 1.28 | 1.01 | 1.14 |
| Lewiston | ID | 0.94 | 0.95 | 0.94 | Reseda Park | CA | 1.28 | 1.01 | 1.14 |
| Boise | ID | 0.83 | 0.89 | 0.86 | Woodland Hills | CA | 1.28 | 1.01 | 1.14 |
| Kellogg | ID | 1.00 | 0.95 | 0.97 | Van Nuys | CA | 1.28 | 1.01 | 1.14 |
| Salt Lake City | UT | 0.84 | 0.92 | 0.88 | San Diego | CA | 1.16 | 1.07 | 1.11 |
| Ogden | UT | 0.84 | 0.95 | 0.89 | Palm Springs | CA | 1.19 | 1.02 | 1.10 |
| Provo | UT | 0.82 | 0.95 | 0.88 | Barstow | CA | 1.18 | 1.02 | 1.10 |
| Cedar City | UT | 0.81 | 0.91 | 0.86 | San Bernardino | CA | 1.17 | 0.98 | 1.07 |
| Phoenix | AZ | 0.88 | 1.02 | 0.95 | Orange County | CA | 1.25 | 1.00 | 1.13 |
| Casa Grande | AZ | 0.84 | 0.86 | 0.85 | Santa Barbara | CA | 1.19 | 1.08 | 1.13 |
| Yuma | AZ | 0.78 | 1.17 | 0.97 | Visalia | CA | 1.12 | 1.00 | 1.06 |
| Douglas | AZ | 0.77 | 0.96 | 0.86 | Bakersfield | CA | 1.11 | 0.98 | 1.04 |
| Tucson | AZ | 0.83 | 0.92 | 0.87 | San Luis Obispo | CA | 1.18 | 1.04 | 1.11 |
| Flagstaff | AZ | 0.90 | 1.07 | 0.98 | Santa Maria | CA | 1.18 | 1.08 | 1.13 |
| Prescott | AZ | 0.93 | 1.02 | 0.97 | Bishop | CA | 1.15 | 1.06 | 1.10 |
| Kingman | AZ | 0.54 | 1.04 | 0.79 | Fresno | CA | 1.14 | 1.13 | 1.13 |
| Albuquerque | NM | 0.78 | 0.94 | 0.86 | Salinas | CA | 1.24 | 1.08 | 1.16 |
| Gallup | NM | 0.79 | 0.99 | 0.89 | Monterey | CA | 1.24 | 1.07 | 1.15 |
| Farmington | NM | 0.77 | 1.09 | 0.93 | El Granada | CA | 1.39 | 1.01 | 1.20 |
| Santa Fe | NM | 0.73 | 0.99 | 0.86 | Pacifica | CA | 1.39 | 1.01 | 1.20 |
| Las Cruces | NM | 0.59 | 1.06 | 0.82 | Redwood City | CA | 1.29 | 1.02 | 1.15 |
| Clovis | NM | 0.74 | 1.07 | 0.90 | San Bruno | CA | 1.29 | 1.02 | 1.15 |
| Roswell | NM | 0.74 | 1.02 | 0.88 | San Francisco | CA | 1.39 | 1.01 | 1.20 |
| Hobbs | NM | 0.68 | 1.14 | 0.91 | Fairfield | CA | 1.25 | 1.01 | 1.13 |
| LasVegas | NV | 1.10 | 0.96 | 1.03 | Newark | CA | 1.39 | 1.01 | 1.20 |
| Reno | NV | 1.07 | 1.12 | 1.09 | Rutherford | CA | 1.25 | 1.01 | 1.13 |
| Elko | NV | 1.05 | 1.11 | 1.08 | Saint Helena | CA | 1.39 | 1.01 | 1.13 |
| Los Angeles | CA | 1.28 | 1.01 | 1.14 | Oakland | CA | 1.25 | 1.01 | 1.20 |
| Santa Monica | CA | 1.20 | 1.01 | 1.10 | Bodega | CA | 1.25 | 1.01 | 1.13 |
| Glendale | CA | 1.22 | 0.99 | 1.10 | San Jose | CA | 1.27 | 1.07 | 1.17 |

| City | | | | | City | | | | |
|---|---|---|---|---|---|---|---|---|---|
| Stockton | CA | 1.17 | 1.03 | 1.10 | Yakima | WA | 0.97 | 0.92 | 0.94 |
| Santa Rosa | CA | 1.25 | 1.01 | 1.13 | Spokane | WA | 1.00 | 0.84 | 0.92 |
| Eureka | CA | 1.11 | 1.02 | 1.06 | Pasco | WA | 1.02 | 0.89 | 0.95 |
| Dixon | CA | 1.19 | 0.98 | 1.08 | Richland | WA | 1.02 | 0.92 | 0.97 |
| Elmira | CA | 1.25 | 1.01 | 1.13 | Anchorage | AK | 1.39 | 1.45 | 1.42 |
| Placerville | CA | 1.19 | 0.98 | 1.08 | Fairbanks | AK | 1.42 | 1.45 | 1.43 |
| Vacaville | CA | 1.25 | 1.01 | 1.13 | Juneau | AK | 1.38 | 1.43 | 1.40 |
| Winters | CA | 1.19 | 0.98 | 1.08 | Ottawa | ONT | 0.96 | 1.10 | 1.03 |
| Sacramento | CA | 1.19 | 0.98 | 1.08 | Toronto | ONT | 0.99 | 1.09 | 1.04 |
| Marysville | CA | 1.19 | 0.98 | 1.08 | Windsor | ONT | 0.97 | 1.10 | 1.03 |
| Chico | CA | 1.12 | 0.98 | 1.05 | Vancouver | BC | 1.12 | 1.08 | 1.10 |
| Redding | CA | 1.12 | 0.98 | 1.05 | Winnipeg | MAN | 0.90 | 1.13 | 1.01 |
| Susanville | CA | 1.15 | 1.01 | 1.08 | Regina | SASK | 0.92 | 1.06 | 0.99 |
| Honolulu | HI | 1.08 | 1.30 | 1.19 | Quebec City | QB | 0.88 | 1.08 | 0.98 |
| The Dalles | OR | 1.01 | 0.94 | 0.97 | Montreal | QB | 0.88 | 1.08 | 0.98 |
| Astoria | OR | 1.01 | 0.96 | 0.98 | Halifax | NS | 0.72 | 1.12 | 0.92 |
| Portland | OR | 1.01 | 0.95 | 0.98 | Edmonton | AB | 0.98 | 1.09 | 1.03 |
| Salem | OR | 1.00 | 1.02 | 1.01 | Brandon | MAN | 0.90 | 1.11 | 1.00 |
| Eugene | OR | 0.98 | 0.94 | 0.96 | Calgary | AB | 0.99 | 1.06 | 1.02 |
| Medford | OR | 0.84 | 1.01 | 0.92 | Darthmouth | NS | 0.71 | 1.10 | 0.90 |
| Bend | OR | 0.90 | 1.01 | 0.95 | Drummondville | QB | 0.88 | 1.08 | 0.98 |
| Pendleton | OR | 0.80 | 1.07 | 0.93 | Fredericton | NB | 0.78 | 1.09 | 0.93 |
| Bellvue | WA | 1.08 | 1.01 | 1.04 | Guelph | ONT | 0.96 | 1.09 | 1.02 |
| Renton | WA | 1.08 | 1.01 | 1.04 | Hamilton | ONT | 0.97 | 1.08 | 1.02 |
| Seattle | WA | 1.08 | 1.01 | 1.04 | Hull | QB | 0.88 | 1.08 | 0.98 |
| Everett | WA | 1.08 | 1.01 | 1.04 | Kamloops | BC | 1.12 | 1.08 | 1.10 |
| Baring | WA | 1.08 | 1.01 | 1.04 | Kelowna | BC | 1.12 | 1.08 | 1.10 |
| Bellingham | WA | 0.95 | 0.95 | 0.95 | Kingston | ONT | 0.95 | 1.10 | 1.02 |
| Index | WA | 1.08 | 1.01 | 1.04 | Kitchener | ONT | 0.96 | 1.10 | 1.03 |
| Bremerton | WA | 1.04 | 1.02 | 1.03 | Lancaster | NB | 0.78 | 1.09 | 0.93 |
| Tacoma | WA | 1.09 | 0.98 | 1.03 | Lethbridge | AB | 0.98 | 1.09 | 1.03 |
| Aberdeen | WA | 1.05 | 1.01 | 1.03 | Lindsay | ONT | 0.98 | 1.10 | 1.04 |
| Vancouver | WA | 1.03 | 0.98 | 1.00 | London | ONT | 0.97 | 1.10 | 1.03 |
| Wenatchee | WA | 0.96 | 1.03 | 0.99 | Moncton | NB | 0.78 | 1.09 | 0.93 |

TABLE 2.2.   Adjustment Indices (Continued)

| Location | | Labor | Material | General |
|---|---|---|---|---|
| Peterborough | ONT | 0.98 | 1.10 | 1.04 |
| Saint Catharines | ONT | 0.98 | 1.09 | 1.03 |
| Saint John | NB | 0.78 | 1.09 | 0.93 |
| Saskatoon | SASK | 0.92 | 1.06 | 0.99 |
| Sherbrooks | QB | 0.88 | 1.08 | 0.98 |
| Thunder Bay | ONT | 0.97 | 1.10 | 1.03 |
| Troms Rivers | QB | 0.88 | 1.06 | 0.97 |
| Victoria | BC | 1.12 | 1.08 | 1.10 |
| Waterloo | ONT | 0.96 | 1.10 | 1.03 |

# 3

# How to Start the Estimate

## GENERAL

This book was written to cover the costs for *new* construction of any size within the confines of the figures and tables herein. These sizes range from residential-size equipment to single boilers of 6 million BTU/hr and single chillers of 500 tons. While it is impossible to cover every piece of equipment which might be used on a project, we hope that 95% of the time you can use this book without reference to other texts. Remodeling, retrofitting, and demolition were not specifically addressed.

Before starting the estimate, certain input is needed and some estimating or calculating must be accomplished in order to size the building's mechanical components. Once these things are accomplished, tentative heating, ventilating, and air conditioning (HVAC) systems must be decided upon and conceptual design decisions must be made in order to use this book. As the design progresses and more design parameters become known, more calculations can be made, and thus a second and better estimate. Actually, we feel that the original budget estimate should be updated twice during the design phase for any project.

The conceptual mechanical systems design need not be carried out on formal drawings to make an estimate. Depending on the size of the project, it can be anything from visualizing the design in your head, to freehand pencil sketches, to hard-line preliminary drawings and specifications.

## THINGS NEEDED

Before starting an estimate, we need architectural plans of the building. The plans should show the proposed plumbing fixtures for us to be able to estimate the plumbing system costs. Ideally, we should have plan and elevation drawings complete enough and firm enough so that we can calculate heating and cooling loads for the building. If this is not possible, we could use rule-of-thumb numbers to estimate these loads.

We should also have a site plan with the building located on the site and complete enough for us to make a determination of the mechanical systems needed for site drainage.

We need to know the project location (so that we can determine the codes in force) and the proposed timetable for design and construction.

## THINGS TO DETERMINE

We must locate the utilities and determine their availability to serve our building. This would include water supply and its pressure, sanitary and storm sewer locations and depths, as well as gas main location if natural gas is the fuel source. If there is no natural gas, the fuel source must be determined.

We also need to determine what codes are in force so that we can add the costs required to meet these codes in our estimate.

## THINGS TO CALCULATE OR ESTIMATE

In order to arrive at a total cost estimate, we need our best guess at system and component sizes. To arrive at these, we need the

Heating load, including the building conduction losses, outdoor air loads for ventilation, and any process or other heat load.

Cooling load, including the building cooling gains and outdoor air loads for ventilation.

Air quantities required for cooling, exhaust systems, and outdoor air for ventilation. In the event the heating system is to be a furnace or furnaces, we need the air quantities for heating.

Water quantities required for heating, for cooling, and for the cooling tower system.

Domestic hot water load.

Building water service size.

Building sewer size.

Building storm sewer size.

## DECISIONS NEEDED

We need to make a tentative selection of the HVAC system. In some cases we may want to compare the costs of two or more systems using this estimating method. We also need to select the component equipment types. The estimating sheets will allow us to compare costs of different types of components if we so choose.

We also need to decide on a method to heat the domestic water.

The plumbing fixtures shown on the architectural plans will probably not be complete. We must determine the need for floor drains, area drains, sump pumps, sewage pumps, hydrants, grease traps, interceptors, drinking fountains, etc., that the architect has not addressed.

We must decide on temperature control methods, monitoring and programming of systems, as well as life safety requirements. Fire protection and detection systems that will be included must be determined. Finally, any special systems which are to be included should be noted.

Do not forget to note any of the principles from the preceding chapter that will affect the cost.

# Chapter

# 4

# Sample Estimate

We have provided a hypothetical building to help us do a sample estimate. Figures 4.1 through 4.4 show us the preliminary plans of the building. Table 4.1 shows the data to calculate the building heating and cooling loads and the resulting loads needed for our estimate.

## Site work estimate

### Water

Starting with the water service, we size the flow for flush-valve water closets at 97 gpm and choose a 2-in service and meter. We know the pressure is 65 psi, suitable for our building without reducing or boosting the pressure. We find that the water company wants the meter inside the building, so we assume we can be provided a meter room on the east side of the south stair. We will run the water line 5 ft south of the building wall. The distance from the water main to the building wall is 130 ft.

From Table 5.1 we add the following:

| | | |
|---|---|---|
| 2-in buried pipe, 130 ft × $11.15/ft = | | $1,449.50 |
| Tap main | | 550.00 |
| Meter (install inside building) | | 1,050.00 |
| | Total cost | $3,049.50 |
| | Use | $3,050.00 |

### Sewer

From experience we know that a 4-in sewer will be adequate. From the main to 5 ft from the building wall is 135 ft. The sewer pipe will be vitrified clay pipe at 6 ft deep, for a cost, from Table 5.2, of $13.60/ft.

135 ft × $13.60/ft = $1,836.00

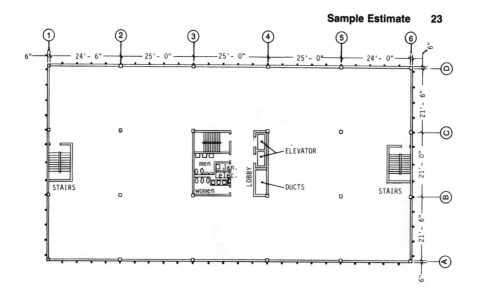

**Figure 4.1**   Typical floor plan (three stories).

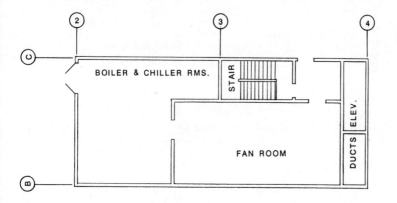

**Figure 4.2**   Penthouse plan.

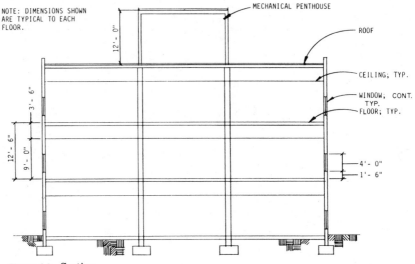

**Figure 4.3**  Section.

From 5 ft outside the building to the plumbing stack riser is measured at 33 ft. This pipe will be cast iron pipe at a cost, from Table 5.2, of $16/ft

$$33 \text{ ft} \times \$16.00/\text{ft} = \$\ \ 528.00$$
$$\text{Total cost}\ \ \overline{\$2,364.00}$$

### Storm sewer

Using the National Plumbing Code and the local code, we find we must size the roof drains on the basis of 4 in of rainfall per hour.

We will collect the drains above the third-floor ceiling and drop down to below the floor in the plumbing chase between the toilet rooms next to the electrical closet. The size to handle 8,125 ft² of roof is 6 in. From this point we have 33 ft of cast iron pipe to 5 ft outside the building. From Table 5.2, at 6 ft deep, the cost of cast iron pipe is

$$33 \text{ ft} \times \$22.55/\text{ft} = \$744.00$$

The distance from 5 ft outside the building wall to the main is 120 ft. This pipe will be vitrified clay. From Table 5.2, at 6 ft deep, the cost of clay pipe is

$$120 \text{ ft} \times \$17.85/\text{ft} = \$2,142.00$$
$$\text{Total cost}\ \ \overline{\$2,886.00}$$

### Cleanouts

Priced at $100 each we have one cleanout in the sewer and one in the storm sewer for a cost of $200.

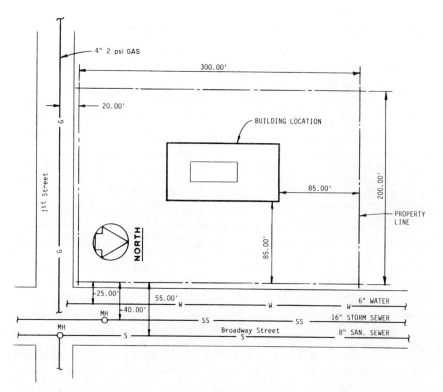

**Figure 4.4** Site plan.

## Gas

Our loads total 659,239 Btu/hr. This is a total of

| | |
|---|---|
| Boiler load | 617,589 Btu/hr |
| Domestic hot-water load (water heater output) | 41,650 Btu/hr |

The heating value of natural gas in Denver is 840 Btu/ft³. The total load is the output required; assuming 80% combustion efficiency, the input is

$$\frac{659,239 \text{ Btu/hr}}{.80} = 824,049 \text{ Btu/hr}$$

The amount of gas required then is

$$\frac{824,049 \text{ Btu/hr}}{840 \text{ Btu/ft}^3} = 981 \text{ cfh}$$

For possible expansion we multiply this by 120% for a gas load of 1,177 cfh.

TABLE 4.1   Sample Building Load Calculations

| Building Description Report |
| --- |

Building location: (1) Denver, Colorado
Design outside temperature for heating: 1°F (default)

| Building Default Parameters | |
| --- | --- |
| General | Lighting |
| Room height, ft: 9 | Watts/square foot: 2 |
| Construction type: Light | % room load contribution: 75 |
| Heating design room temperature, °F: 72 | Daily lighting schedule, hr: 10 |
| Cooling design room temperature, °F: 75 | Schedule starting at: Override |
| Heating supply air temperature, °F: 110 | Room classification: A |
| Cooling supply air temperature, °F: 55 | |
| Infiltration, 1 air change/hour | |
| People | Equipment and miscellaneous |
| Square feet/person: 100 | Watts/ft$^2$: 0.25 |
| Daily occupancy, hr: 10 | % room load contribution: 100 |
| Occupancy starting at: Override | Daily equipment operation: 10 |
| Sensible Btu's/hour/person: 255 | Operation starting at: Override |
| Latent Btu's/hour/person: 255 | |

| Building Energy Code Check | |
| --- | --- |

Denver, Colorado
Degrees latitude: 39.1
Heating degree days: 6,016                Exterior design temperature, °F: 91.0
Solar factor: 126.4                        Interior design temperature, °F: 78.0
                                           Building stores: 1

A. Building surfaces

| Roofs: | ID | Name | U value | Area, ft$^2$ | |
| --- | --- | --- | --- | --- | --- |
| | R1 | 2 in concrete | 0.05 | 6,316 | |

| Walls: | ID | Name | U value | Area, ft$^2$ | Delta T |
| --- | --- | --- | --- | --- | --- |
| | W1 | Inside panel | 0.091 | 6,930 | 44 |

| Glass: | ID | Name | U value | Area, ft$^2$ | Shading coefficient |
| --- | --- | --- | --- | --- | --- |
| | G1 | Reflective | 0.55 | 3,510 | 0.50 |

| Floors: | ID | Name | U value | Area | |
| --- | --- | --- | --- | --- | --- |
| | F1 | Slab | —Floor not over air space— | | |

B. Summation of surfaces

| | Area, ft$^2$ | Average U value | Required U value |
| --- | --- | --- | --- |
| Roofs: | 6,316 | 0.050 | 0.076 |
| Walls: | 10,440 | 0.245 | 0.275 (includes glass) |
| Floors: | 0 | 0.000 | 0.080 |

TABLE 4.1    Sample Building Load Calculations (Continued)

C. Overall values

|  | Actual | Required | Pass |
|---|---|---|---|
| U values: | 0.172 | 0.200 | Yes |
| Thermal transfer values: | 26.32 | 33.25 | Yes |

### Building Block Load

Floor area: 24,375 ft$^2$
Heating loads:
  Conduction:        Infiltration: 130,665        Total load: 353,227
  Roof:    22,421
  Wall:    44,774
  Glass:  137,065
  Floor:   18,300
Cooling loads: (Month 8, Hour 17)

| Conduction: | Solar load: | 156,795 | Total conduction: | 31,916 |
|---|---|---|---|---|
| Roof: 8,645 | Sensible people: | 61,200 | Total sensible: | 381,001 |
| Wall: 10,342 | Latent people: | 61,200 | | |
| Glass: 12,928 | Lighting: | 112,362 | Room, cfm: | 23,644 |
| Floor: 0 | Equipment and miscellaneous: | 18,727 | | |

Cooling return air pickup:
  Roof:                                          2,881  Total sensible return air load: 421,337
  Lighting:                                    37,454
  Equipment and miscellaneous:          0

### Room Load Summary, cfm

| 1 SE | 434 | 2 SE | 434 | 3 SE | 436 |
|---|---|---|---|---|---|
| 1 NE | 347 | 2 NE | 347 | 3 NE | 353 |
| 1 NW | 414 | 2 NW | 414 | 3 NW | 432 |
| 1 SW | 468 | 2 SW | 468 | 3 NW | 485 |
| 1 N | 355 | 2 N | 355 | 3 N | 392 |
| 1 W | 2,141 | 2 W | 2,141 | 3 W | 2,264 |
| 1 S | 659 | 2 S | 659 | 3 S | 660 |
| 1 E | 1,896 | 2 E | 1,896 | 3 E | 1,931 |
| 1 C | 1,314 | 2 C | 1,314 | 3 C | 1,469 |
| | 8,028 | | 8,028 | | 8,422 |
| | | | | | 24,478 |

Diversity: 23,644/24,478 = 96%

NOTE: N = north, W = west, S = south, E = east, C = center.

Sizing the pipe at 1.0-in total pressure loss for a distance of 150 equivalent feet to the meter, we reach a size of 2 in. From the main to the building wall is 110 ft. From Table 5.4, we can calculate the cost of gas service as

110 ft × $10.75/ft = $1,182

**Exhibit 4.1**
**SITE WORK SUMMARY SHEET**

|  | Total cost, $ |
|---|---|
| Water service | 3,050 |
| Sanitary sewer | 2,364 |
| Storm sewer | 2,886 |
| Cleanouts | 200 |
| Manholes | — |
| Gas service | 1,182 |
| Miscellaneous    None | |
| Portion of estimate, this section | 9,682 |

Total these costs on Exhibit 4.1, Site Work Summary Sheet, for total site work cost.

### Plumbing Estimate

From our site work estimate we have 65 psi water pressure. Using Exhibit 4.2, Plumbing Cost Estimating Summary Sheet, as a checklist for our estimate, we see that for a three-story building we do not need pressure reducing or pressure boosting.

The water heater can be sized from *ASHRAE Systems*, 1984, page 34.14. From ASHRAE Table 7,

$$
\begin{array}{lll}
18 \text{ lavatories} \times 2 \text{ gph} & = 36 & \text{gph} \\
3 \text{ service sinks} \times 20 \text{ gph} & = 60 & \text{gph} \\
\hline
\text{Total} & 96 & \text{gph}
\end{array}
$$

Demand factor:    $.3 \times 96 \text{ gph} = 28.8 \text{ gph}$

Storage factor:    $2 \times 28.8 \text{ gal} = 57.6 \text{ gal}$

From Table 6.2 we will pick the 75-50 gas water heater which has a tank size of 75 gal and a recovery rate at sea level of 50 gph.

Since our building floor is at or about grade level, we will have no need for a sewage or sump pump.

We summarize the plumbing fixtures as follows:

| | |
|---|---|
| Electric water coolers | 3 |
| Service sinks | 3 |
| Lavatories, wall-hung | 9 |
| Lavatories, countertop | 9 |
| Urinals | 6 |

Wall closets, flush-valve, wall-hung                                          15

We will put a 2-in floor drain in each toilet and one in each
janitor's closet, total                                                        9

We also decide to put two floor sinks in the chiller room, two
in the boiler room, and one in the fan room, all 3 in.                         5

From the National Plumbing Code we will use one 2-in roof
drain on the penthouse roof that will be drained to the main
roof.                                                                          1

The main roof requires two 4-in roof drains, each with a 5-in
horizontal line to one common 5-in riser in the plumbing wall
next to the electrical closet. Site work picks up the drain from
there.                                                                         2

Roof drain piping length is

  Horizontal 5-in, approximately 70 ft − 30 ft                        = 40 ft
  (15 ft included with each drain price) Vertical 5-in                = 35 ft

We will design one freezeproof hydrant on each side of the
building, total                                                                4

## Water mains

The cold-water main is sized as follows:*

  15 water closets (flush-valve) × 10 f.u. = 150 f.u.
  6 urinals            × 5 f.u. =   30 f.u.
  18 lavatories      × 2 f.u. =   36 f.u.
  3 janitor sinks    × 4 f.u. =   12 f.u.
  3 electric water coolers  × ½ f.u. =   1½ f.u.
                             229 f.u. = 97 gpm

We use a 2-in water line and meter with a flow of 97 gpm.
The hot-water main is similarly sized.*

  18 lavatories   × 1½ f.u.      = 27 f.u.
  3 janitor sinks × 3 f.u.      =  9 f.u.
                         36 f.u. = 23 gpm

Flow of 23 gpm requires a 1-in water line.

    The cold-water-main length from the meter room on the east side of
the south stair to the plumbing wall behind the fixtures is

2-in horizontal piping, approximately                                          37 ft

2-in vertical piping from ground floor through third floor = 25
ft + 2 ft above the floor                                                      27 ft

The length of the 1-in hot-water main from the top of the
boiler room to the first floor is 47 ft − 2 ft above the floor                 45 ft

---

* *ASHRAE Systems*, 1970.

Sewer main lengths are

| | |
|---|---|
| No-hub sewer vertical riser, 4 in from roof of penthouse to first floor | 48 ft |
| Galvanized steel vent from first-floor ceiling space to penthouse roof, estimate size at $2\frac{1}{2}$ in | 40 ft |

We see no miscellaneous costs, so we are ready to price the plumbing.

From Table 6.2 the water heater assembly cost for a 75-50 heater is $1,215.

From Table 6.4

| | | |
|---|---|---|
| Electric water coolers | 3 × $750 = | $2,250 |
| Service sinks | 3 × 975 = | 2,925 |
| Lavatories, wall-hung | 9 × 575 = | 5,175 |
| Lavatories, countertop | 9 × 600 = | 5,400 |
| Urinals | 6 × 625 = | 3,750 |
| Water closets, wall-hung | 15 × 700 = | 10,500 |
| Toilet room drains | 9 × 225 = | 2,025 |
| Remote floor sinks | 5 × 575 = | 2,875 |
| Roof drains   2-in | 1 × 325 = | 325 |
| 4-in | 2 × 425 = | 850 |
| Overflow drains (2-in drain is scuppered to roof) | | |
| 4-in | 2 × 325 = | 650 |
| Wall hydrants, nonfreeze | 4 × 240 = | 960 |

From Table 6.6

| | | |
|---|---|---|
| Roof drain piping 5-in horizontal galvanized steel 40 ft × $35.15/ft | = | 1,406 |
| 5-in vertical galvanized steel 35 ft × $31.60/ft | = | 1,106 |
| Galvanized vent, $2\frac{1}{2}$ in vertical 40 ft × $14.90/ft | = | 596 |
| No-hub sewer, 4-in vertical 40 ft × $11.95/ft | = | 574 |

From Table 6.5

2-in copper, horizontal
    37 ft. × $11.25/ft                                        =    416
2-in copper, vertical, with tees each floor
    27 ft × $10.20/ft                                         =    275
1-in copper, vertical, with tees each floor
    45 ft × $6.65/ft                                          =    299

Entering all costs on our summary sheet, Exhibit 4.2, we have a total plumbing cost, before profit, of $43,572.

## HVAC Estimate

Table 4.1 shows the data used to calculate the heating and cooling loads for the building and the resulting loads. We will use the following summary sheets as checklists for our conceptual design and estimating:

Heating (Exhibit 4.3)

Electric heating (Exhibit 4.4)

Cooling generation and distribution (Exhibit 4.5)

Cooling, heating, and HVAC equipment and piping (Exhibit 4.6)

Air-moving equipment (Exhibit 4.7)

Air-distribution equipment (Exhibit 4.8)

Before starting an estimate we must make tentative decisions on the types of systems and equipment to use. Surveying the building loads will help us in this effort. The following items are all covered in Chapter 7.

### Heat generation

Our total heating load would occur at night with the air system off and the outdoor air dampers closed; we would have conduction losses *and* infiltration losses totaling 353,227 Btu/hr.

Coil load is 264,362 Btu/hr for morning warm-up. (See Heating Coils, this section.) So the total load on the boiler is 617,589 Btu/hr. We can use a cast iron boiler with either an atmospheric or a forced-draft burner.

### Heating pumps

Radiation pump, size at $20°dt$ ($dt$ is the difference between outside and inside temperatures) is

**Exhibit 4.2**
**PLUMBING COST ESTIMATING SUMMARY SHEET**

|  |  |  | Total cost, $ |
|---|---|---|---|
| Water pressure-reducing valve assembly |  |  | — |
| Water pressure-booster systems |  |  | — |
| Water heater assembly |  |  | $1,215 |
| Hot-water generator assembly |  |  | — |
| Sewage pump assemblies |  |  | — |
| Sump pump assemblies |  |  | — |
| Drinking fountains | No. of units ___ × $/unit ___ | = |  |
| Electric water coolers | No. of units _3_ × $/unit _750_ | = | 2,250 |
| Service sinks | No. of units _3_ × $/unit _975_ | = | 2,925 |
| Lavatories, wall-hung | No. of units _9_ × $/unit _575_ | = | 5,175 |
| Lavatories, countertop | No. of units _9_ × $/unit _600_ | = | 5,400 |
| Kitchen sinks, single-compartment | No. of units ___ × $/unit ___ | = | — |
| Kitchen sinks, double-compartment | No. of units ___ × $/unit ___ | = | — |
| Disposers | No. of units ___ × $/unit ___ | = | — |
| Wash fountains | No. of units ___ × $/unit ___ | = | — |
| Urinals | No. of units _6_ × $/unit _625_ | = | 3,750 |
| Water closets, tank-type | No. of units ___ × $/unit ___ | = | — |
| Water closets, flush-valve (wall-hung) | No. of units _15_ × $/unit _700_ | = | 10,500 |
| Bathtubs with shower | No. of units ___ × $/unit ___ | = |  |
| Shower stalls | No. of units ___ × $/unit ___ | = |  |
| Integral bath and shower | No. of units ___ × $/unit ___ | = |  |
| Toilet room drains |  |  | 2,025 |
| Remote drains |  |  | — |
| Remote floor sinks |  |  | 2,875 |
| Roof drains |  |  | 1,825 |
| Wall hydrants |  |  | 960 |
| Roof drain piping |  |  | 2,512 |
| Water mains |  |  | 990 |
| Galvanized steel mains |  |  | 596 |
| Cast iron no-hub mains |  |  | 574 |
| Miscellaneous _None_ |  |  |  |
| Portion of estimate, this section |  |  | 43,572 |

$$\frac{353,227}{500 \times 20°dt} = 35.3 \text{ gpm}$$

Coil pump, size at $20°dt$ is

$$\frac{264,362}{500 \times 20°dt} = 26.49 \text{ gpm}$$

**Building heating**

Heating can be accomplished with baseboard radiation zoned with a room thermostat and automatic valve for each of the four corners and four sides of the building. Water should be scheduled from outdoor temperature with a pump for the system.

We have 380 lineal feet of wall per floor, with three floors, for a total of 1,140 lineal feet. The 354,000-Btu/hr load divided by 1,140 lineal feet = 310 Btu/ft capacity required. Using 75% of the available wall space gives a radiation capacity range required of 415 Btu/(hr · ft), well within the range of baseboard radiation.

**Air supply**

We have a minimum of nine zones per floor because of orientation and exposure. These are the four corners, four sides, and the core. To handle these zones plus any other zones caused by occupancy or internal load differences, we decide on a variable-air-volume (VAV) air system using chilled-water coils with hot-water coils for morning warm-up. From past experience, it appears that a one-fan system (no return air fan) will meet our needs, since we will have very little pressure drop on the return air side. The 23,644-cfm block load is used to size the fan system and is within the range of factory-packaged equipment.

We select a horizontal HVAC unit with combination mixing box and filters at 23,644 cfm, and we approximate the static pressure from past experience at $4\frac{1}{2}$ to 5-in water column (wc).

**Heating coils**

We have several trial conditions to investigate to set the load on these coils: (1) amount of outdoor air required for proper ventilation; (2) amount of outdoor air required for mixed air temperature for winter-time "free cooling"; and (3) amount of heat required for morning warm-up. Calculations are as follows:

1. The building has 8,125 ft$^2$ per floor × 3 floors = 24,375 ft$^2$. With an occupancy load of 200 ft$^2$ per person, the occupancy is

$$\frac{24{,}375}{200} = 122 \text{ people}$$

Ventilation outdoor air required is 15 cfm per person. Therefore for 122 people, we calculate the total ventilation outdoor air required as

$$122 \times 15 \text{ cfm/person} = 1{,}830 \text{ cfm}$$

2. The fan handles 23,644 cfm; to provide 55° supply air we have the following:
72°F room
1°F outdoor air temperature
71°$dt$

$$\text{Percent outdoor air} = \frac{dt \text{ [room to leaving air temperature (LAT)]}}{dt \text{ [room to outside air temperature (OAT)]}}$$

$$\text{Percent outdoor air} = \frac{17}{71} = 24\%$$

Air quantity = $23{,}644 \times .24 = 5{,}674$ cfm

This second quantity is the larger of the two figures in numbers 1 and 2, so we will use 5,600 cfm.
The amount of outdoor air taken in will be controlled by the air-handling unit mixed air temperature controller. With *no* fixed minimum on heating there will be no outdoor air load on the boiler.

3. Amount of heat required for morning warm-up. Assume the reduced night temperature to be 60°F with a 1-hr warm-up period to warm the air to 72°. The fan might supply the room cfm load summary (Table 4.1) when in the morning warm-up cycle.

$$24{,}478 \text{ cfm} \times 12°dt \times .9 \text{ (constant at Denver)} = 264{,}362 \text{ Btu/hr}$$

We believe that one row of heating coils will handle this load and we will base our estimate on this premise. We will use a pumped coil arrangement with a three-way mixing valve for control.

### Entry heating

The entry will be heated by a floor-mounted electric cabinet heater.

### Cooling generation plant

We have decided on a chilled-water cooling system. The chiller would be sized based on the total sensible air load on the coils plus the outdoor air load imposed by the minimum outdoor air setting of the summer controls. Calculating this load we have the following:

1. The minimum outdoor air required for ventilation for occupants equals 1,830 cfm (from heating calculations). Exhaust of toilet rooms, plus janitor's and electrical closets equals

   Area = 170 ft$^2$/floor $\times$ 2 cfm/ft$^2$ $\times$ 3 floors = 1,020 cfm exhaust

   Setting the minimum outdoor air at 15% of the fan capacity gives us

   23,644 cfm $\times$ .15 = 3,547 cfm

   enough to provide the exhaust plus some pressurization of the building. We will use 3,547 cfm outdoor air:

   3,547 $\times$ (91° − 75°)$dt$ $\times$ .9 = 51,077 Btu/hr

2. The cooling load on the coils and compressor is

   Total sensible load = 421,337 Btu/hr

   Outdoor air load = $\underline{\quad 51,077 \text{ Btu/hr}}$
                         472,414 Btu/hr

   $$\frac{472,144 \text{ Btu/hr}}{12,000 \text{ Btu/ton}} = 39.3 \text{ tons}$$

This is within the range of reciprocating chillers. We will use a factory-packaged reciprocating chiller with an air-cooled condenser located on the roof outside the equipment room.

**Cooling coils and pumps**

The coil load is the same as the compressor load. We believe that it will take a six-row coil and will use this for our estimate.

We want constant flow through the chiller, so we will pump the chiller with a three-way bypass valve on the cooling coil for control. Since both the coil and the chiller are subject to freezing, we will fill the system with a 20% glycol solution.

The pump is to be sized at 2.4 gpm/ton:

   40 tons (approximately) $\times$ 2.4 = 96 gpm

Using these calculations we can estimate the heating and cooling equipment and piping costs.

**Using the heating cost estimating summary sheet** (see Exhibit 4.3)

**Boiler assembly.** From Figure 7.1, a cast iron boiler assembly with atmospheric burner rated at 620 MBH (1,000 Btu/hr) costs $14.60/MBH:

620 MBH × $14.60/MBH = $9,052

**Combustion air heater assembly**

620 MBH × 16.26 factor = 10,000 Btu/hr

From Figure 7.10,

$$\frac{10{,}000 \text{ Btu/hr}}{3410 \text{ Btu/(hr} \cdot \text{kW)}} = 2.93 \text{ kW}$$

Cost for a horizontally mounted unit heater at 3 kW = $29.50/MBH.

10 MBH × $29.50/MBH = $295

**Heating pump assemblies.** The radiation pump must provide 35.3 gpm at an estimated head of 38 ft. From Figure 7.2, curve 5, the cost of a series 60 in-line pump assembly rated at 40 gpm is

40 gpm × $41.50/gpm = $1,660

The coil pump must supply 26.4 gpm at an estimated head of less than 18 ft. From Figure 7.2, curve 4, the series 60 in-line pump assembly rated at 27 gpm costs

27 gpm × $45/gpm = $1,215

**Pot feeder assembly.** One 2-gal pot feeder (see Chapter 7 text) costs $523.

**Distribution mains for the perimeter heating system.** We can feed the first and second floors from the first-floor ceiling. This constitutes approximately two-thirds of the 35.4 gpm load:

35.4 gpm × .67 = 23.7 gpm    or    237,000 Btu/hr

From curve 2, Figure 7.5, we extrapolate to $14/MBH:

237 MBH × $14/MBH = $3,318

We can feed the third floor from the second-floor ceiling; this is approximately one-third of the load:

324 MBH − 237 MBH = 87 MBH

Curve 1, Figure 7.5, ends at 140 MBH; using this point, the cost is $27.50/MBH;

$$87 \text{ MBH} \times \$27.50/\text{MBH} = \$2,392$$

**Transmission Mains.**  The cost would be based on the distance from the boiler room wall to the duct chase and down to the first-floor ceiling.
Horizontal piping length of approximately 30 ft × 2 pipes = 60 ft. From Chapter 7, at 35.3 gpm, 2-in pipe is required.
Vertical 2-in pipe from overhead in the penthouse to the second-floor ceiling is measured at 23 ft:

$$23 \text{ ft} \times 2 \text{ pipes} = 46 \text{ ft}$$

Vertical pipe from the second-floor ceiling to the first-floor ceiling is measured at 12 ft × 2 pipes = 24 ft. From Chapter 7, at 23.7 gpm, $1\frac{1}{2}$-in pipe is required.
Costs from Table 7.1 are:

Horizontal 2-in pipe, no fittings, hangers for insulated pipe:

$$60 \text{ ft} \times \$9.63/\text{ft} = \$ 578$$

Vertical 2-in pipe, no fittings, supports each floor:

$$46 \text{ ft} \times \$7.18/\text{ft} = \$ 330$$

Vertical $1\frac{1}{2}$ in pipe, no fittings, supports each floor:

$$24 \text{ ft} \times \$5.74/\text{ft} = \$ \ \underline{138}$$
$$\text{Total} = \overline{\$1,046}$$

**Radiation assemblies.**    Heat loss equals 354,000 Btu/hr. From Table 7.4 the Slimline VK-AR-14 has 640 Btu/ft capacity:

$$\frac{354,000 \text{ Btu}}{640 \text{ Btu/ft}} = 553 \text{ ft required}$$

$$553 \times 1.2 \text{ factor from text} = 664 \text{ estimated footage}$$

$$664 \text{ ft} \times \$58.83/\text{ft} = \$39,063$$

**Electric heaters.**    The entry heater is estimated at about 10 kW. From Figure 7.11, curve 2F, a floor-mounted entry heater costs $37/MBH:

$$10 \text{ kW} \times 3.41 \text{ MBH/kW} \times \$37/\text{MBH} = \$1,262$$

**Heating coil assemblies.**    We have tentatively picked a factory-packaged air-handling unit. Table 7.13 shows the sizes and cubic feet per minute selection. We will use a size 50.

Note that a two-coil arrangement is standard with this unit. Coils cost $1,333.

**Heating coil piping assemblies.**  We have two coils of equal height; the flow rate to each is

$$\frac{26.4 \text{ gpm}}{2} = 13.2 \text{ gpm}$$

From Table 7.17, under "three-way mixing valve for heating," column 5, the cost for $1\frac{1}{4}$-in piping to one coil is

$950

To this we add the cost for the second coil, from column 6,

$$\begin{array}{rl} & \$1300 \\ \text{Total} & \overline{\$2250} \end{array}$$

**Teflon pump connectors.**  One for each coil inlet and outlet of the $1\frac{1}{2}$-in size, from Table 7.1b:

4 × $141 = $564

This ends our heating portion estimate, and we total these costs on the summary sheets, Exhibits 4.3 and 4.4.

We now calculate our cooling generation and distribution costs.

**Chiller plant**

Our load is 39.3 tons; from Figure 7.13, curve 4, we have a cost of approximately

$610/ton × 40 tons = $24,400

**Chilled water pump**

Capacity needed is 96 gpm at 38 ft of estimated head. From Figure 7.2, curve 5, the Bell & Gossett series 60 in-line pump is cheaper than the base-mounted pumps. Using the needed flow and costs per unit of flow, we obtain

96 gpm × $29/gpm = $2,784  for the cost of the pump

## Exhibit 4.3
## HEATING COST ESTIMATING SUMMARY SHEET

| | | Total cost, $ |
|---|---|---|
| Boiler plant, type  Cast iron, atmospheric | | |
| 620 MBH | @ 14.60 $/MBH = | 9,052 |
| Combustion air heater, type Electric cabinet heater (Horizontal) | | |
| 10 MBH | @ 29.50 $/MHB = | 295 |
| Heating pump(s) series  60 | | |
| No. of pumps 1 @ 40 gpm × 41.50 $/gpm = | | 1,660 |
| Pot feeder assembly  2 gal | 5 gal = | 523 |
| Antifreeze feeder assembly, field-erected factory-mounted | = | — |
| Heating transmission mains | | 1,046 |
| Perimeter heating distribution mains | | 5,710 |
| Heating coil assemblies | | 1,333 |
| Heating coil piping assemblies | | 2,250 |
| Heating coil flexible connectors | | 564 |
| Heating coil pumps | | 1,215 |
| Primary-secondary bridles | | — |
| Reheat coils and piping assemblies | | — |
| Unit heaters: | | |
| Type S | | — |
| Type P | | — |
| Cabinet heaters: | | |
| Type B | | — |
| Type D | | — |
| Convectors: | | |
| Type W | | — |
| Type FG | | — |
| Type SG | | — |
| Radiation: | | |
| Slimline style, 664 ft | @ 58.33 $/ft = | 39,063 |
| style, ft | @ $/ft = | |
| Radiant panels | | |
| style, ft | @ $/each = | |
| style, ft | @ $/each = | |
| Miscellaneous None | = | — |
| Heating portion estimate, this sheet | = | 62,711 |

## Antifreeze feeder assembly

From Chapter 7, the cost of field-erected assembly is $1,450. We can enter these loads and total them on the Cooling Generation and Distribution Summary Sheet, Exhibit 4.5.

We are now ready to proceed with our cooling, heating, and HVAC equipment and piping costs.

Exhibit 4.4
ELECTRIC HEATING COST ESTIMATING SUMMARY SHEET

|  | Total cost, $ |
|---|---|
| Electric unit heaters: | |
| Horizontal mounting | — |
| Vertical mounting | — |
| Electric cabinet heaters: | |
| Floor-mounted | 1,262 |
| Ceiling-mounted | — |
| Electric radiation: | |
| Residential style | — |
| Commercial-institutional: | |
| Style CSH | — |
| Style DSH | — |
| Electric radiant ceiling panels | — |
| Specialty heaters: | |
| Kickspace heaters | — |
| Floor heaters | — |
| Wall heaters | — |
| Miscellaneous  None | — |
| Electric heating portion estimate, this sheet | 1,262 |

Exhibit 4.5
COOLING GENERATION AND DISTRIBUTION COST ESTIMATING
SUMMARY SHEET

|  | Total cost, $ |
|---|---|
| Chiller plant, type  air-cooled, reciprocating | |
| 40  tons @  610  $/ton  = | 24,400 |
| Air-cooled condenser assembly _____ tons × _____ $/ton = | — |
| Chilled water pump(s),  series 60 | |
| No. of pumps  1  @  96  gpm ×  29  $/gpm  = | 2,784 |
| Pot feeder assembly    2 gal    5 gal    = | — |
| Antifreeze feeder assembly field-erected, factory-mounted  = | 1,450 |
| Cooling tower assembly, type _____ | |
| _____ tons    @    _____ $/ton = | — |
| Chemical treatment assembly, type _____  = | — |
| Tower freeze protection, size _____ tons = | — |
| Tower water filtering assembly    Total cost  = | — |
| Condenser water pump(s), series _____ | |
| No. of pumps _____ @ _____ gpm × _____ $/gpm = | — |
| Condenser water mains | |
| _____ ft    @    _____ $/ft  = | — |
| Miscellaneous  None    = | — |
| Cooling portion estimate, this sheet    = | 28,634 |

**Chilled water transmission mains.** From the discussion in Chapter 7 under Chiller Room Piping, it appears that we do not need to add additional piping to reach from the chiller to the air-handling unit.

**Air conditioning unit assembly.** We have previously selected a size 50 unit. From Figure 7.24, curve 3, for $4\frac{1}{2}$-in static pressure, the cost at 23,644 cfm is $.47/cfm.

Unit assembly cost = 23,644 × $.47 = $11,113

From Table 7.11 we select the following options:

4-in permanent filters: No extra cost.

Inlet vanes:

$$23,644 \text{ cfm} \times \$.05/\text{cfm} = \$1,182$$

From Chapter 7 text, we see that arrangement 3 costs $500 to $1,000 more. We believe that arrangement 3 best suits our application.

Add              $ 1,000
      Total   $13,295

**Cooling coil assemblies.** We have previously seen that the size 50 unit has two coils of equal size:

$$\frac{96 \text{ gpm}}{2} = 48 \text{ gpm each}$$

From Table 7.13, the coil cost for six-row cooling coils is $4,480.

**Cooling coil piping assemblies.** From Table 7.17, column 2, "three-way bypass valve for cooling," we find a piping cost for the first coil of

                                              $1,450
Adding for two coils high, we have            $1,690
                           Total cost    $3,140

**Teflon pump connectors.** We will use one for each coil inlet and outlet. For the 2-in size, Table 7.1*b*,

4 × $154 = $616

**Exhibit 4.6**
**COOLING, HEATING, AND HVAC EQUIPMENT AND PIPING COST ESTIMATING**
**SUMMARY SHEET**

|  | Total cost, $ |
|---|---|
| Fan coil assemblies | — |
| Unit ventilator assemblies | — |
| Transmission mains:<br>Chilled water<br>Heating, water (enter on heating summary<br>sheet, Exhibit 4.3) | — |
| Perimeter distribution piping mains:<br>Chilled water<br>Heating water (enter on heating summary sheet, Exhibit 4.3)<br>Drain pan piping (see plumbing, Exhibit 4.2) | — |
| Heating and ventilating unit assemblies | — |
| Air conditioning unit assemblies | 13,295 |
| Heating and ventilating multizone assemblies | — |
| HVAC multizone assemblies:<br>Two-deck<br>Three-deck<br>Double-duct | —<br>—<br>— |
| Cooling coil assemblies* | 4,480 |
| Chilled water coil piping assemblies* | 3,140 |
| Chilled water coil flexible connectors* | 616 |
| Chilled water coil pumps* | — |
| Chilled water primary-secondary bridles* | — |
| Miscellaneous___None_____ | — |
| Cooling portion estimate, this sheet | 21,531 |

* Enter heating components on Exhibit 7.4

   This ends our cooling portion, and we can now enter these costs on the summary sheet, Exhibit 4.6.

   We are now ready to proceed to the air-moving equipment costs (see Chapter 9).

### Exhaust fan for toilets

We previously saw that at 2 cfm/ft$^2$ of toilet area the air quantity for the fan was 1,020 cfm. From Figure 9.3, curve 1, the cost for dome-type roof exhausters is $.57 cfm at 1,000 cfm:

   1,020 cfm × $.57/cfm = $581

**Exhibit 4.7**
**AIR-MOVING EQUIPMENT COST ESTIMATING SUMMARY SHEET**

|  | Total cost, $ |
|---|---|
| Centrifugal fans | — |
| Inlet vanes and controls | — |
| Axial fans | — |
| Inlet vanes and controls | — |
| Cabinet fans | — |
| Inlet vanes and controls | — |
| Vaneaxial fans (controls included) | — |
| Dome-type roof exhaust fans | 581 |
| Dome-type wall exhaust fans | — |
| Utility fans | — |
| Coating | — |
| Laboratory upblast exhaust fans | — |
| Kitchen hood upblast exhaust fans | — |
| Propeller fans (small D.D.) | — |
| Propeller fans (large B.D.) | — |
| Residential exhaust fans: | |
| Hoods | — |
| Toilets | — |
| Miscellaneous___None_____ | — |
| Portion of estimate, this sheet | 581 |

This is all the air-moving equipment we have for this summary sheet, Exhibit 4.7.

We now calculate our air distribution equipment costs.

**Sound attenuators**

From Chapter 9 text, 7-ft-long attenuators (on the supply and return air sides of the fan) cost $.09/cfm.

23,644 cfm × 2 × $.09/cfm = $4,256

**Toilet exhaust ductwork**

From Figure 9.5, for a three-story building, the cost of the ductwork is $1.53/cfm:

1,020 cfm × $1.53 = $1,561

**Variable-air-volume (VAV) ductwork** (Table 9.3)

For supply ducts to VAV units, costs are

23,644 cfm (block load) × $1.45/cfm     = $34,284

**Return air duct** (Table 9.3)

23,644 cfm × $.25/cfm                    = $ 5,911

**Outdoor air duct** (Table 9.3)

It appears that this duct could extend directly through the roof to an intake hood and that $.25/cfm is too high a cost, so use $.10/cfm:

23,644 cfm × $.10                        = $ 2,364
Total VAV ductwork to VAV units          $42,559

**Supply ducts from VAV units to diffusers** (Table 9.3)

24,478 cfm (room load summary) × $.50/cfm = $5,911

**Intake hood**

Figure 9.7, curve 3, shows a cost at 10,000 cfm of $.07/cfm. This is extrapolated at this same cost:

23,644 cfm × $.07/cfm = $1,655

**Relief hood**

We will need a relief hood when the system modulates to take outdoor air for cooling.

From Figure 9.7, curve 2, for the exhaust hood with backdraft dampers, the cost is $.10/cfm at 10,000 cfm. This is also extrapolated at the same cost:

(23,644 − 1,020)cfm × $.10 = $2,262

**VAV control units**

By sketching VAV unit locations and zones we arrived at 18 VAV units per floor. This number is subject to the designer's desires and can differ from this. By using 18 units per floor times three floors we have 54 VAV units. We determine the average load per unit as follows:

$$\frac{24,478 \text{ cfm (room load summary)}}{54 \text{ units}} = 453 \text{ cfm average}$$

From Figure 9.8, curve 1, for shutoff units with pneumatic control, the cost at 453 cfm per unit is $.50/cfm. The total cost is

24,478 cfm × $.50/cfm = $12,239

### Diffusers, return grilles, and exhaust registers

By sketching diffuser locations we arrived at 54 locations per floor or 151 cfm of load each. Estimating from Figure 9.9, we will use the Trane two-slot diffuser, curve 10. At 151 cfm each, the cost is $.30/cfm. The total cost is

24,478 cfm × $.30/cfm = $7,343

We will use 16 return air grilles per floor. The load for each is determined as follows:

$$\frac{24,478}{16 \times 3 \text{ floors}} = 509 \text{ cfm each}$$

From Figure 9.9, curve 8, for square perforated-face ceiling grilles, the cost is approximately $.06/cfm. The total cost is

24,478 cfm − 1,020 cfm exhaust = 23,458 cfm × $.06/cfm = $1,407

### Exhaust registers

We have four per floor times three floors for a total of 12. The average load for each is

$$\frac{1,020 \text{ cfm}}{12} = 85 \text{ cfm}$$

From Figure 9.9, curve 7, at 125 cfm each, the cost is $.23/cfm. Using this cost, the total cost is

1,020 cfm × $.23/cfm = $235

Summing each of these costs gives a total cost for grilles, registers, and diffusers of $8,985. We can now enter these costs on the summary sheet, Exhibit 4.8.

### Temperature control estimate

We will estimate the control cost from Figure 10.1 as a percentage of the costs of Chapter 7,

$114,138

**Exhibit 4.8**
**AIR DISTRIBUTION EQUIPMENT COST ESTIMATING SUMMARY SHEET**

|  | Total cost, $ |
|---|---|
| Built-up filter banks | — |
| Prefabricated plenums | — |
| Sound attenuators | 4,256 |
| Ductwork: Single-zone supply air, return air, outside air: All systems | |
| Toilet exhaust: All systems | 1,561 |
| VAV supply air to boxes, return air and outside air: All systems | 42,559 |
| VAV supply air from boxes: All systems | 12,239 |
| Hoods: Intake | 1,665 |
| Exhaust | 2,262 |
| Louvers: Intake | — |
| Exhaust | — |
| Dampers: Manual opposed blade | — |
| Counterbalanced | — |
| Automatic | — |
| Fire | — |
| VAV control units: Shutoff type | 12,239 |
| Fan-powered type | — |
| Dual-duct type | — |
| Shutoff with reheat | — |
| Fan powered with reheat | — |
| Grilles, registers, diffusers: All types | 8,985 |
| Miscellaneous  None | — |
| Portion of estimate, this sheet | 85,756 |

plus Chapter 9,

$$\begin{array}{ll} & \underline{\$\ 86,337} \\ \text{Total} & \$200,475 \end{array}$$

Curves 2 and 3 in Figure 10.1 show the lower and upper limits of control prices for projects researched. We choose 17% for our estimate:

$200,475 \times 17\% = \$34,081$

**Balancing cost estimate**

We will estimate the balancing cost from Figure 11.1 as a percentage of the costs of Chapters 7 and 9, i.e., $200,475. We will use 3%:

$200,475 × 3% = $6,014

**Insulation cost estimate**

We will estimate the insulation cost from Figure 12.1 as a percentage of the costs of Chapter 6,

$  43,572

and Chapter 7,

$114,138

Total  $157,710

Curves 1 and 2 show the lower and upper limits of insulation contracts. We choose 9% for our estimate:

157,710 × 9% = $14,194

We can now enter all our costs on the cost estimating summary sheet in Exhibit 4.9 (see Chapter 15).

**Cost Estimating Summary** (Exhibit 4.9)

**General**

We will use our cost estimating summary sheets along with Figure 15.1 as checklists to summarize and complete our estimate.

**Cost summary**

Having entered all our costs on the summary sheets we decide on the multipliers. Our building is small, the owner is ready to start, and we anticipate 6 months of construction time:

The area index for Denver is 1.0.

No inflation is anticipated, and the factor is 1.0.

The building is straightforward with no design unknowns, and the factor for design contingency is 1.0.

From Figure 15.1 the profit factors are as follows. Site work plus plumbing plus HVAC, all to be done by the mechanical contractor, with one profit number on the total:

Exhibit 4.9
COST ESTIMATING SUMMARY SHEET

| Portion of estimate | Cost from summary sheets, $ | Multipliers | | | | | Cost to mechanical contractor, $ | Mechanical contractor's markup multiplier | Cost estimate to general contractor, $ |
|---|---|---|---|---|---|---|---|---|---|
| | | Area index | Inflation factor | Design contingency | Profit, each contractor, % | Total multiplier | | | |
| Site work, Chapter 5 | 9,682 | 1.0 | 1.0 | 1.0 | 1.11 | 1.11 | 10,747 | 1.0 | 10,747 |
| Plumbing, Chapter 6 | 43,572 | 1.0 | 1.0 | 1.0 | 1.11 | 1.11 | 48,365 | 1.0 | 48,365 |
| Heating and cooling, Chapter 7 | 114,138 | 1.0 | 1.0 | 1.0 | 1.11 | 1.11 | 126,693 | 1.0 | 126,693 |
| Packaged equipment, Chapter 8 | None | | | | | | — | | |
| Air distribution, Chapter 9 | 86,337 | 1.0 | 1.0 | 1.0 | 1.115 | 1.115 | 96,266 | 1.0 | 96,266 |
| Controls, Chapter 10 | 34,081 | 1.0 | 1.0 | 1.0 | 1.112 | 1.112 | 38,171 | 1.05 | 40,080 |
| Balancing, Chapter 11 | 6,014 | 1.0 | 1.0 | 1.0 | 1.125 | 1.125 | 6,766 | 1.05 | 7,104 |
| Insulation, Chapter 12 | 14,194 | 1.0 | 1.0 | 1.0 | 1.12 | 1.12 | 15,897 | 1.05 | 16,692 |
| Fire protection, Chapter 13 | None | | | | | | | | |
| Special systems, Chapter 14 | None | | | | | | | | — |
| Other items | None | | | | | | | | — |
| Bonds | None | | | | | | | | — |
| Total cost estimate of the mechanical systems | | | | | | | | | 345,947 |

$ 9,682
$ 43,572
$114,138

Profit factor, 1.11%.

$ 9,682 × 1.11    Price = $ 10,747

$ 43,572 × 1.11    Price = $ 48,365

$114,138 × 1.11    Price = $126,693

We assume that air moving and distribution will be a subcontract or that if done in-house by the mechanical contractor, a separate profit factor will be employed. Profit factor on

$86,337 = 1.115    Price = $96,266

The profit on the controls cost of

$34,081 = 1.12    Price = $38,171

The profit on the balancing cost of

$6,014 = 1.125    Price = $6,766

The profit on the insulation cost of

$14,194 = 1.12    Price = $15,897

The mechanical contractor's markup is addressed next. The bidding market is very favorable; there is not much work, so we will apply only 5% to the controls, balancing, and insulation contracts:

Controls    $38,171 × 1.05 = $40,080

Balancing    $ 6,766 × 1.05 = $ 7,104

Insulation    $15,897 × 1.05 = $16,692

No other items are to be included. Bonds will not be required because we will preselect bidders. The total for our estimate, therefore, is

$345,947.

## Check figures

Our estimate to the owner then is $346,000. Our check figures on the cost per square foot and percentage of total cost are

$$\frac{\$346,000}{24,375 \text{ ft}^2} = \$14.19/\text{ft}^2$$

The owner's estimate of total building cost was $70/\text{ft}^2$. We know from our past experience that the mechanical cost should be 19 to 22% of the total:

$$\frac{\$14.19}{\$70/\text{ft}^2} = 20.3\%$$

These check figures both appear reasonable, and the estimate is completed.

# 5

# Site Work

Site work costs and utility costs were based on pipe sizes as noted. Depths of burial that vary from those used change the price. For shallower trenches the estimator may use $.60 per lineal foot for the cost of excavation, backfill, and compaction for each foot of depth difference. Street traffic affects the street cut costs and should always be analyzed. The street cut costs herein are based on average city streets with asphalt paving; costs associated with major arteries, highways, etc., increase these costs. If landscaping must be replaced, this adds to the cost. Type of soil also affects these costs. Costs shown here are based on trenches dug in average soil with no shoring. Any time shoring is needed for loose soil or deep trenches the cost increases. Trenches were 2 ft wide with a sand bed and sand to surround the pipe and with backfill air tamped in 12 in lifts. Exhibit 5.1 is the site work summary sheet.

### Water Service

Water service costs were based on trench depths to provide 4 ft of cover over the pipe. Pipe used for pricing was type K copper, using 95-5 solder for pipe sizes 3 in and under. For 4- and 6-in pipe we used ductile iron pipe.

Costs shown in Table 5.1 are based on Denver, Colorado, standards. Meter vault is to be added only if the meter is installed outside the building; other costs are additive. Meter assembly is essentially the same cost if installed inside the building.

Figure 5.1 a and b shows the assemblies upon which the costs were based.

**Exhibit 5.1**
**SITE WORK SUMMARY SHEET**

|  | Total cost, $ |
|---|---|
| Water service | _____ |
| Sanitary sewer | _____ |
| Storm sewer | _____ |
| Cleanouts | _____ |
| Manholes | _____ |
| Gas service | _____ |
| Miscellaneous | _____ |
| _____ | |
| _____ | |
| _____ | |
| Portion of estimate, this section | _____ |

TABLE 5.1    Water Service Costs

| Pipe size, in | Pipe material | Buried pipe, $/ft | Tap main, saddle, valve, and valve box, $ | Meter with 3-valve bypass, $ | Meter vault and manhole cover, $ |
|---|---|---|---|---|---|
| ¾ | Copper | 7.20 | 320 | 400 | 225 |
| 1 | Copper | 7.65 | 330 | 500 | 225 |
| 1½ | Copper | 9.00 | 430 | 750 | 500 |
| 2 | Copper | 11.15 | 550 | 1050 | 500 |
| 3 | Copper | 16.25 | 975 | 2,100* | 675 |
| 4 | Ductile iron | 14.15 | 2,150 | 3,025* | 775 |
| 6 | Ductile iron | 16.00 | 3,025 | 5,700* | 1,075 |

* Meters are compound meters.

## Sewers

Costs shown in Table 5.2 are for sanitary and storm sewers. Costs are shown for 6- and 8-ft-deep trenches and for pipe sizes of 4, 6, and 8 in. Prices were for gasketed vitrified clay pipe.

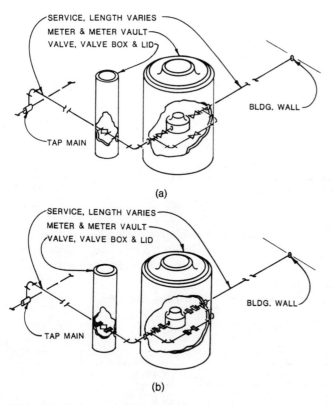

Figure 5.1   Water service. (*a*) Copper pipe; (*b*) iron pipe.

## Cleanouts and Manholes

Cleanouts in pipe outside the building can be priced at $100 each. This includes the cost of the wye and one-eighth bend in the pipe, the riser, and the cleanout with a cast iron cover set in a 2-ft-square concrete pad.

Manhole costs were estimated using precast concrete rings and 24-in-diameter, 400-lb, cast iron frame and cover. Table 5.3 shows the

**TABLE 5.2   Sewer Piping Costs**

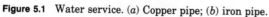

| Pipe size, in | 6-ft-deep trench, $/lin ft | | 8-ft-deep trench, $/lin ft | |
|---|---|---|---|---|
| | Vitrified clay | Cast iron | Vitrified clay | Cast iron |
| 4 | 13.60 | 16.00 | 14.70 | 17.10 |
| 6 | 17.85 | 22.55 | 18.90 | 23.60 |
| 8 | 22.85 | 31.00 | 23.90 | 32.09 |

costs for manholes of different sizes and depths. Excavation and backfill are included.

TABLE 5.3   Manhole Costs

| Manhole diameter, ft | Manhole depth, ft | Cost, $ |
|---|---|---|
| 4 | 6 | 805 |
| 4 | 8 | 1,050 |
| 5 | 8 | 1,300 |
| 5 | 10 | 1,525 |

## Gas Service

Prices were for steel pipe, coated, wrapped, and buried 2-ft deep. Table 5.4 shows these costs for pipe sizes 1 in through 6 in. Note that in many areas the utility company installs the service and the meter and bills the building owner directly. Include this cost only if the contractor installs the gas service. The meter is owned and set by the utility company.

TABLE 5.4   Gas Service Costs

| Gas pipe size, in | Cost, $/lin ft |
|---|---|
| 1 | 7.50 |
| 2 | 10.75 |
| 3 | 15.25 |
| 4 | 19.75 |
| 6 | 36.75 |

# 6

# Plumbing

The work covered in this chapter consists of plumbing equipment, fixtures, drains, and pipe mains to fixture locations. By adding the appropriate costs of these components, an accurate estimate can be obtained. Exhibit 6.1 is the plumbing cost estimating summary sheet.

Water piping used for pricing was type L copper pipe using 95-5 solder with copper fittings and sweat end valves where available. Waste and vent piping used for pricing was cast iron no-hub and galvanized steel with cast iron drainage fittings.

## Plumbing Equipment

### Water-pressure-reducing-valve assembly

Figure 6.1a shows the assembly used for pricing. Up through a 2-in pipe size we used a one-pressure-reducing-valve (PRV) station; for 3-in pipe size we used a three-PRV station. PRVs were sized using 5-psi pressure variation and 25 to 50 psi reduction in pressure.

Figure 6.1b shows the cost of the station in dollars per gpm.

### Water-pressure-booster system

The booster system priced was the Bell & Gossett 70M pumping system, with pumps operating at 3500 rpm, 208/60/3 current. The system included a steel base and factory-mounted pump(s), motor starter(s), a manual alternator, PRVs, shutoff valves, galvanized steel piping, no-flow shutdown feature, and all standard controls shown in Bell & Gossett's catalog. Costs are shown in Table 6.1 for several water

Exhibit 6.1
**PLUMBING COST ESTIMATING SUMMARY SHEET**

| | | Total cost, $ |
|---|---|---|
| Water pressure reducing valve assembly | | _____ |
| Water pressure booster systems | | _____ |
| Water heater assembly | | _____ |
| Hot water generator assembly | | _____ |
| Sewage pump assemblies | | _____ |
| Sump pump assemblies | | _____ |
| Drinking fountains | No. of units _____ × $/unit _____ = | _____ |
| Electric water coolers | No. of units _____ × $/unit _____ = | _____ |
| Service sinks | No. of units _____ × $/unit _____ = | _____ |
| Lavatories, wall-hung | No. of units _____ × $/unit _____ = | _____ |
| Lavatories, counter-top | No. of units _____ × $/unit _____ = | _____ |
| Kitchen sinks, single-compartment | No. of units _____ × $/unit _____ = | _____ |
| Kitchen sinks, double-compartment | No. of units _____ × $/unit _____ = | _____ |
| Disposers | No. of units _____ × $/unit _____ = | _____ |
| Wash fountains | No. of units _____ × $/unit _____ = | _____ |
| Urinals | No. of units _____ × $/unit _____ = | _____ |
| Water closets, tank-type | No. of units _____ × $/unit _____ = | _____ |
| Water Closets, flush valve | No. of units _____ × $/unit _____ = | _____ |
| Bathtubs with shower | No. of units _____ × $/unit _____ = | _____ |
| Shower stalls | No. of units _____ × $/unit _____ = | _____ |
| Integral bath and shower | No. of units _____ × $/unit _____ = | _____ |
| Toilet room drains | | _____ |
| Remote drains | | _____ |
| Remote floor sinks | | _____ |
| Roof drains | | _____ |
| Wall hydrants | | _____ |
| Roof drain piping | | _____ |
| Water mains | | _____ |
| Galvanized steel mains | | _____ |
| Cast iron no-hub mains | | _____ |
| Miscellaneous_____ | | _____ |
| Portion of estimate, this section | | _____ |

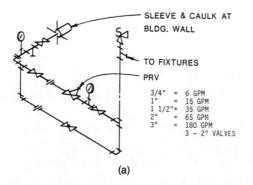

(a)

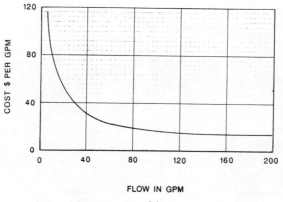

FLOW IN GPM

(b)

**Figure 6.1** (*a*) Water pressure reducing valve assembly; (*b*) water pressure reducing valve assembly cost.

**TABLE 6.1 Water Pressure Booster System Cost**

| Pump assembly size, Bell & Gossett, Model 70M | Total gpm | Head pressure developed, psi | Cost, $ |
|---|---|---|---|
| 1, size 1¼ A5A | 70 | 50 pi | 5,100 |
| 2, size 1¼ A5A | 140 | 50 pi | 10,400 |
| 2, size 1½ A7B | 240 | 50 pi | 11,400 |
| 2, size 1¼ A7A | 140 | 80 pi | 10,400 |
| 2, size 1½ A10B | 240 | 80 pi | 11,600 |
| 1, size 1½ A10B | | | |
| 1, size 2  B20C | 320 | 80 pi | 13,200 |

flows at two different head pressures. These head pressures add to the suction pressure available to the pumping station.

### Water heater assemblies

Figure 6.2 shows the assembly used for pricing gas-fired and electric water heaters. Water piping was sized at $\frac{3}{4}$ in for 30-, 40-, and 50-gal heaters; for the larger sizes of 75-50, 50-75, 75-75, 100-75 and electric 82 gal, we used 1-in water piping. Gas pipes were sized for the gas input to the heater. Tanks priced were all glass-lined. Breeching was galvanized steel duct, 10 ft long.

**TABLE 6.2 Water Heater Costs**

| Heater | | Cost of the |
|---|---|---|
| Model | Size | heater assembly, $ |
| Gas* | 30-30 | 815 |
| | 40-32 | 835 |
| | 50-32 | 875 |
| | 75-50 | 1,215 |
| | 50-75 | 1,400 |
| | 75-75 | 1,585 |
| | 100-75 | 1,925 |
| Electric† | 30 | 600 |
| | 40 | 615 |
| | 52 | 645 |
| | 82 | 815 |

\* The first number is the tank size (such as 30 in 30-30); the second number (also 30 in 30-30) the gas input in thousands of Btu's per hour.
† The number is the tank size. These units were single-element heaters: 5,500 W, 25-gal/hr recovery, 90° F rise.

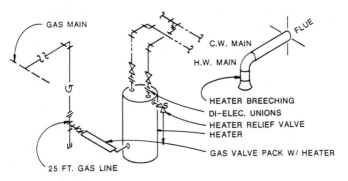

**Figure 6.2**  Water heater assembly.

Table 6.2 shows the costs of these heaters. Add $50 for LP gas heaters.

### Hot-water generator assemblies

Figure 6.3a shows the assembly used for pricing these generators. Tanks used were 125 psi with American Society of Mechanical Engineers (ASME) label. Glass- and copper-lined tanks were priced. The heaters priced were sized for tank recovery in 1 hr with 200°F boiler water heating domestic water from 40 to 140°F.

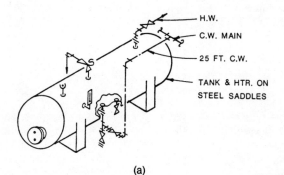

(a)

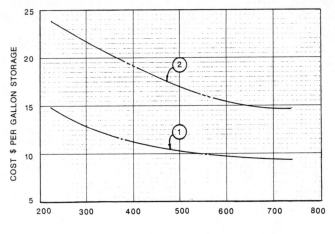

STORAGE TANK SIZE IN GALLONS

(b)

**Figure 6.3** (a) Hot water generator assembly; (b) hot water generator assembly cost: (1) glass-lined tank; (2) copper-lined tank.

Figure 6.3*b* shows the cost of these generator assemblies in dollars per gallon of storage capacity.

### Recirculating pump assemblies

For pricing refer to Chapter 7, Heating and Cooling, under Pump Assemblies.

**TABLE 6.3 Sewage and Sump Pump Costs**

| Pump description | Tank diameter, in | Flow each pump, gpm | Cost with no tank, $ | Cost with steel tank, $ |
|---|---|---|---|---|
| Sewage: | | | | |
| Simplex, $\frac{1}{2}$ hp | 30 | 70 | 3,000 | 3,670 |
| Simplex, $\frac{3}{4}$ hp | 30 | 120 | 3,100 | 3,770 |
| Simplex, 1 hp | 30 | 200 | 3,950 | 4,620 |
| Duplex, $\frac{1}{2}$ hp each | 42 | 70 | 5,925 | 6,860 |
| Duplex, $\frac{3}{4}$ hp each | 42 | 120 | 6,125 | 7,060 |
| Duplex, 1 hp each | 48 | 200 | 7,175 | 8,110 |
| Sump pump: | | | | |
| $1\frac{1}{4}$ in, $\frac{1}{3}$ hp | 30 | | 880 | 1,535 |
| 2 in, $\frac{1}{2}$ hp | 30 | | 1,175 | 1,830 |

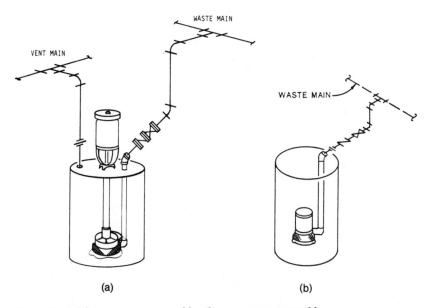

(a)                                    (b)

**Figure 6.4**   (*a*) Sewage pump assembly; (*b*) sump pump assembly.

## Sewage and sump pump assemblies

Figure 6.4a shows the assembly used for pricing sewage pumps. This figure shows a simplex pump installation. For a duplex installation, each pump would pump into a wye in a common line to the main.

Figure 6.4b shows the assembly used for pricing small sump pumps.

Table 6.3 shows the installed costs of sewage and sump pump assemblies with prices including a coated steel tank and prices without the tank. The pumps are mounted on a steel-plate tank cover. Tanks for pumps were 72-in deep. Pumps were chosen at 15-ft head. Starters and automatic alternators were included.

## Plumbing Fixtures and Drains

All fixtures were priced using medium-cost trim, cast brass P traps with cleanouts, and chromium-plated supplies with stops. Wall-hung lavatories were priced with wall hanger and urinals, and wall-hung water closets were priced with carriers.

Figures 6.5 through 6.14 show the fixture assemblies used for pricing.

Table 6.4 shows the costs of fixtures, drains, and hydrants based on the assemblies shown. Assemblies show the fixture standing alone. This obviously changes with the configuration of the fixture layout in

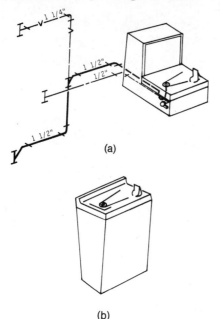

(a)

(b)

**Figure 6.5** (a) Drinking fountain or (b) electric water cooler assembly.

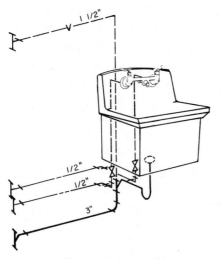

**Figure 6.6** Service sink assembly.

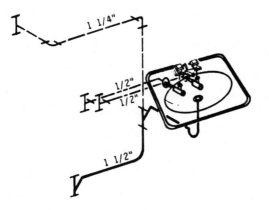

**(a)**

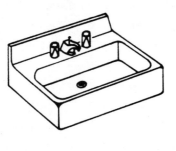

**(b)**

**Figure 6.7** Lavatory assembly. (*a*) Countertop; (*b*) wall-hung.

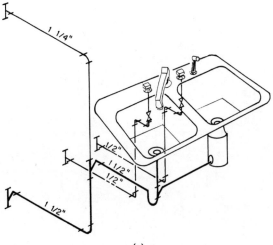

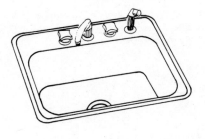

(a)

(b)

**Figure 6.8** Kitchen sink assembly. (*a*) Two-compartment; (*b*) single-compartment.

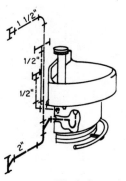

**Figure 6.9** Wash fountain assembly.

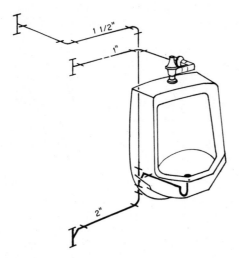

**Figure 6.10** Urinal assembly.

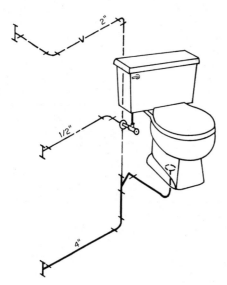

**Figure 6.11** Tank-type water closet assembly.

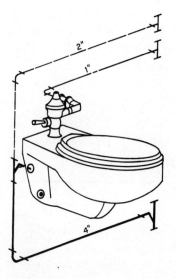

**Figure 6.12** Flush-valve water closet assembly.

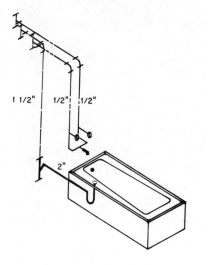

**Figure 6.13** Bathtub and shower assembly.

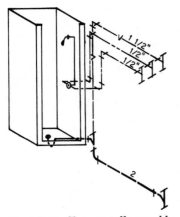

**Figure 6.14** Shower stall assembly.

**TABLE 6.4  Fixtures and Drains Costs**

| Item and description | Cost, $ |
| --- | --- |
| Drinking fountain, VC* | 550 |
| Electric water cooler, 8 gal/hr | 750 |
| Service sink, enameled CI,† wall-mounted | 975 |
| Lavatory, VC, wall-hung | 575 |
| Lavatory, VC, countertop | 600 |
| Kitchen sink, enameled CI, 24 × 21 single | 700 |
| Kitchen sink, enameled CI, 32 × 21 double | 750 |
| For stainless steel sink, add $120 | |
| Disposer, $\frac{1}{2}$ hp in sink | 175 |
| Wash fountain, 54-in semicircular precast | 1650 |
| Urinal, VC, wall-hung siphon-jet | 625 |
| Water closet, VC tank-type | 575 |
| Water closet, VC wall-hung flush valve | 700 |
| Bathtub with shower head, 5-ft enameled CI | 825 |
| Shower Stall, 30″ enameled steel and molded stone | 800 |
| Integral bath and shower package, molded plastic | 800 |
| Toilet room drain, 2 in with 15-ft pipe | 225 |
| Remote drain, 3 in with 50-ft pipe | 425 |
| Remote floor sink, 3 in with 50-ft pipe | 575 |
| Roof drains, 15-ft pipe to main or wall discharge: | |
| 2 in | 325 |
| 3 in | 375 |
| 4 in | 425 |
| Wall hydrants: | |
| Bronze hydrant, not nonfreeze | 200 |
| Bronze hydrant, nonfreeze style | 240 |
| Bronze hydrant, nonfreeze in bronze case | 280 |
| Overflow roof drain next to main roof drain, with 5 ft of pipe and sanitary tee: | |
| 2 in | 250 |
| 3 in | 275 |
| 4 in | 325 |

\* VC = vitreous china.
† CI = cast iron.

the toilet rooms for lavatories, urinals, and water closets in commercial buildings. Back-to-back fixture layouts can also decrease the plumbing costs. There are also many variations on each of the fixtures shown here as well as other types of fixtures not included in this work. It has been our experience that a reasonable conceptual estimate can be made using these prices.

## Pipe Mains

Tables 6.5 and 6.6 show the cost of copper water mains and galvanized steel and cast iron no-hub waste and vent mains. By estimating the footage and size of the mains from the building entry to the user equipment and fixture banks and adding the cost of these mains to the cost of equipment and fixtures, you can complete the estimate.

**TABLE 6.5  Piping Costs: Type L Copper Pipe**

| | Horizontal piping with no fittings: hangers for insulated pipe, $/lin ft | Vertical piping, $/lin ft | |
| Pipe size, in | | With no fittings, supports each floor | With tees each 12 ft; supports each floor |
|---|---|---|---|
| $\frac{1}{2}$ | 5.95 | 3.10 | 4.45 |
| $\frac{3}{4}$ | 6.25 | 3.45 | 5.25 |
| 1 | 7.55 | 4.40 | 6.65 |
| $1\frac{1}{4}$ | 8.05 | 4.95 | 7.45 |
| $1\frac{1}{2}$ | 8.55 | 5.45 | 8.20 |
| 2 | 11.25 | 6.90 | 10.20 |
| $2\frac{1}{2}$ | 13.80 | 9.95 | 15.45 |
| 3 | 16.10 | 12.35 | 19.10 |
| 4 | 24.00 | 17.50 | 26.85 |

**TABLE 6.6  Piping Costs: Standard Weight Galvanized Steel and Standard Weight Cast Iron No-Hub**

| | Horizontal piping with no fittings; hangers for noninsulated pipe | | Vertical piping, $/lin ft | | | |
| | | | With no fittings; supports each floor | | With tees each 12 ft; supports each floor | |
| Pipe size, in | Galvanized steel | Cast iron no-hub | Galvanized steel | Cast iron no-hub | Galvanized steel | Cast iron no-hub |
|---|---|---|---|---|---|---|
| $1\frac{1}{2}$ | 7.80 | — | 5.70 | — | 8.65 | — |
| 2 | 9.60 | 9.35 | 7.15 | 6.90 | 10.95 | 9.15 |
| $2\frac{1}{2}$ | 12.20 | — | 10.25 | — | 14.90 | — |
| 3 | 14.80 | 11.40 | 12.85 | 9.40 | 21.25 | 12.65 |
| 4 | 20.40 | 15.55 | 16.80 | 11.95 | 28.95 | 16.60 |
| 5 | 35.15 | — | 31.60 | — | — | — |
| 6 | 43.40 | 22.10 | 35.75 | 18.25 | — | 25.70 |

# 7

# Heating and Cooling

The heating, cooling, and refrigeration equipment, systems, and unit assemblies priced and discussed in this section generally are installed by the mechanical or plumbing and heating contractor. Included are generating systems, pumping systems, distribution piping systems, and terminal heating and cooling equipment.

All piping 2 in and below was priced as screwed pipe, valves, and fittings; pipe $2\frac{1}{2}$ in and above was priced using welded and flanged construction. Heating piping was sized on $20°dt$ unless noted. Cooling piping was sized on $10°dt$ or 2.4 gpm/ton. Condenser water piping was sized on $10°dt$ or 3 gpm/ton.

## Heating

### Water boiler plants

The curves in Figure 7.1 show the cost in dollars per MBH for natural gas–fired boiler heating plants. All boilers have 30-psi working pressure. Curves are shown for cast iron boilers with atmospheric burners, cast iron boilers with forced-draft burners, and steel fire tube Scotch Marine boilers. For oil-fired or combination gas- and oil-fired boilers, an estimator could arrive at the price for the job by deducting the price of a gas burner and adding the price of the other type of burner. In addition, refer to the oil tank pricing chart under Special Systems (Chapter 14) and add these costs to Exhibit 7.1.

The curves reflect the installed cost of the entire heating generating system in the boiler room except for pumping (see Figures 7.2 and 7.3), chemical feeder assemblies, and combustion air heater (see Unit Heaters, Figure 7.6 or 7.10). Add these costs to the boiler plant cost in Exhibit 7.1. Included in the boiler room pricing for each size of boiler studied are

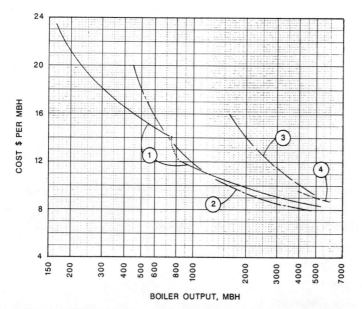

**Figure 7.1** Boiler plants. (1) Cast iron boiler with atmospheric burner; (2) cast iron boiler with forced-draft burner, Weil-McLain series 88; (3) Scotch Marine steel fire tube boiler with forced draft burner; (4) cast iron boiler with forced-draft burner, Weil-McLain series 94.

Boiler and burner

4-in-thick concrete mounting pad

Breeching to flue

Gas piping

Combustion air louvers

Expansion tank and air fitting

Air separation system

Relief valve piped to drain

Boiler water feeder assembly

Boiler room piping

Boiler shutoff valves (two gate valves)

### Boiler and burner standards

*Cast iron boiler with atmospheric burner.*   Prices were obtained from the Peerless Heater Co. for series 61 and 211A models including all standard trim plus the following:

McDonnell-Miller no. 63 low-water cutoff

Factory Mutual Insurance Co. gas train plus a slow-opening

**Exhibit 7.1**
**HEATING COST ESTIMATING SUMMARY SHEET**

| | Total cost, $ |
|---|---|
| Boiler plant, type_____ | |
| _____ MBH          @ _____ $/MBH = | _____ |
| Combustion air heater, type_____ | |
| _____ MBH          @ _____ $/MBH = | _____ |
| Heating pump(s) series _____ | |
| No. of pumps_____ @ _____ gpm × _____ $/gpm = | _____ |
| Pot feeder assembly: 2 gal      5 gal | _____ |
| Antifreeze feeder assembly: field-erect, factory-mounted | _____ |
| Heating transmission mains, cost/ft | _____ |
| Perimeter heating distribution mains | _____ |
| Heating coil assemblies | _____ |
| Heating coil piping assemblies | _____ |
| Heating coil flexible connectors | _____ |
| Heating coil pumps | _____ |
| Primary-secondary bridles | _____ |
| Reheat coils and piping assemblies | _____ |
| Unit heaters: | |
|   Type S | _____ |
|   Type P | _____ |
| Cabinet heaters: | |
|   Type B | _____ |
|   Type D | _____ |
| Convectors: | |
|   Type W | _____ |
|   Type FG | _____ |
|   Type SG | |
| Radiation: | |
| _____ style, _____ ft          @ _____ $/ft = | _____ |
| _____ style, _____ ft          @ _____ $/ft = | _____ |
| Radiant panels: | |
| _____ style, _____ ft          @ _____ $/each = | _____ |
| _____ style, _____ ft          @ _____ $/each = | _____ |
| Miscellaneous _____ | _____ |
| Heating portion estimate, this sheet | _____ |

diaphragm gas valve and electronic flame safety control for units with 400,000-Btu and above output
High-pressure gas cutoff switches for units with 2,500,000-Btu and above input

*Cast iron—forced-draft burner.*  Prices were obtained from Weil-McLain for series 88 and 94 models, including all standard trim plus the following:

Mcdonnell-Miller no. 63 low-water cut-off
Ultraviolet electronic flame safety control
High- and low-pressure gas cutoff switches for units with 2,500,000 Btu input and above
High-low fire-up to 5 million Btu
Modulating fire above 5 million-Btu output

*Scotch Marine fire tube boiler.*  Prices were obtained from Cleaver-Brooks for model CB, including all standard trim plus the following:

Mcdonnell-Miller no. 63 low-water cutoff
High- and low-pressure gas cutoff switches
A second motorized gas safety shutoff valve plus an additional leakage test cock

**Breeching to flue.**  For boilers with atmospheric burners, galvanized steel breeching was used. It was from 22 to 16 gauge in thickness and from 20 to 35 ft in length from the smallest to largest size boilers.

For forced-draft boilers, 10-gauge black steel breeching, 20 ft in length, was the basis of pricing.

**Gas piping.**  The basis of pricing was 50 ft of black steel schedule 40 pipe, sized with a total pressure drop of 0.3 in water gauge (wg).

**Combustion air louver.**  Sized for 15 cfm of air flow for each cubic foot per minute of gas flow through the louver at 100 ft/min face velocity.

**Expansion tank and air fitting.**  Tanks were chosen using average sizes from past projects of varying types with up to three stories of piping above the tank. ASME tanks with sight glass and air fitting were used.

**Air separation system.**  Bell & Gossett (B&G) boiler fittings were used through 256,000 Btu of boiler output. Above that, B&G Rolairtrol air separators (without screen) were priced. Included is the piping to the expansion tank and drain valve with 10 ft of drain piping.

**Boiler water feeder.**  The boiler water feeder assembly consisted of a $\frac{3}{4}$-in vacuum breaker and B&G no. 12 water feeder in a three-valve bypass with a vacuum breaker drain pipe to the drain. While this is plumbing work, it was included here for ease in presentation.

**Boiler room piping.**  Boiler room piping was sized using an average of the lengths involved from past projects studied in each size category. We added 60 ft of piping on the smallest boiler, graduating up to 200 ft of piping for a 5,500-MBH system.

### Combustion air heater assemblies

For each cubic foot of gas burned it takes approximately 15 ft$^3$ of air for combustion. For each 1 cfm of gas flow at 1000 Btu/ft$^3$ × 60 min/hr = 60,000 Btu/hr boiler input:

> 60,000 Btu/hr input × .83 efficiency = 49,800 Btu/hr output

so for each 15 cfm of combustion air we can get 49,800 Btu/hr boiler output. To heat 15 cfm of combustion air from 0 to 50°F, the equation is

> cfm × $dt$ × constant        = Btu/hr
> 15 cfm × (50° − 0°) × 1.08 = 810 Btu/hr

So for each 49,800 Btu/hr (49.8 MBH) boiler output we need 810 Btu/hr to heat the combustion air 50°. This equates to 16.26 Btu per boiler MBH output.

Multiply the boiler output in MBH by 16.26 to get the combustion air heater size in Btu's per hour and refer to unit heater chart, Figure 7.6 or 7.10, for the cost.

For design temperature differentials other than 50°, the heater size is determined as follows:

$$\text{Heater size} = 16.26 \times \frac{\text{new } dt}{50°} \times \text{boiler output MBH}$$

### Pump assemblies

Pumps were priced using B&G and Aurora iron body pumps. All figures are installed costs. From researching past projects in our office, three ranges of head pressures kept reoccurring. For small jobs, 15 to 20 ft of head, for medium-sized jobs, 35 to 40 ft of head, and for cooling and larger heating jobs, 60 to 70 ft of head. Pumps (other than "boosters") were priced on this basis. Figures 7.2 and 7.3 show the costs of pumps in dollars per gallon per minute of flow.

In the booster pumps, series 100, all sizes are the same price. We picked a 1¼-in pump at 15 gpm of flow, which is about the maximum flow for 1¼-in pipe. All associated piping costs were for 1¼-in pipe. In the HV series, they are again all the same cost within that series. We

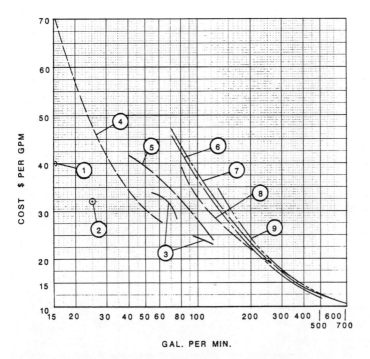

**Figure 7.2**   Pump assemblies. (1) B&G series 100; (2) B&G series HV; (3) B&G booster pumps; (4) B&G series 60, 18-ft head; (5) B&G series 60, 38-ft head; (6) B&G series 1510, 38-ft head; (7) B&G series 1510, 70-ft head; (8) B&G series 80, 38-ft head; (9) B&G series 80, 70-ft head.

picked a $1\frac{1}{2}$-in HV pump at 25 gpm, which again is about the maximum flow for $1\frac{1}{2}$-in pipe. All associated piping costs were for $1\frac{1}{2}$-in pipe. Moving into the booster series we priced each pump at the point shown on the pump performance curves, Figure 7.4. Note that the flow rate *and the head pressure* vary with each pump size, whereas other pumps were priced at common head pressures. This changes the shape of the curve and cost per gallon per minute relationship in a family of pumps.

The second family of pumps priced were series 60 in-line pumps. These are generally small-gallonage pumps. One group was priced at low pressures of 18 ft of head. The second group was priced using 38 ft of head.

Base-mounted end-suction pumps for larger gallonages were priced using series 1510 flexible-coupled single-suction pumps. Two groups were priced, one at 38 ft and one at 70 ft of head. Base-mounted, B&G series 80 in-line pumps were also priced at 38 and at 70 ft of head.

Base-mounted double-suction pumps were priced using Aurora series 410 pumps. Again, two groups were priced, one at 35 and one at 70 ft of head.

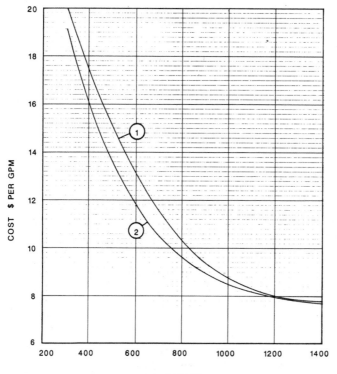

GAL. PER MIN.

**Figure 7.3**  Double-suction pump assemblies. (1) Aurora series 410, 70-ft head; (2) Aurora series 410, 35 ft head.

Pump prices include the motor, which was chosen so that it would be nonoverloading, with the impeller provided. Generally, the pump was selected using the next to the largest standard size impeller shown for that pump. Motors $\frac{1}{3}$ hp and less were single-phase; motors $\frac{1}{2}$ hp and above were three-phase with a magnetic starter. The starter was furnished but not installed. All pumps were selected with 1,750-rpm motors for quiet operation.

In pricing the pump and its associated piping, we have sized the piping using the following maximum flow rates throughout:

| Pipe size, in | Maximum flow, gpm | Pipe size, in | Maximum flow, gpm |
|---|---|---|---|
| $\frac{3}{4}$ | 4 | 3 | 140 |
| 1 | 8 | 4 | 300 |
| $1\frac{1}{4}$ | 16 | 5 | 550 |
| $1\frac{1}{2}$ | 25 | 6 | 940 |
| 2 | 50 | 8 | 2,000 |
| $2\frac{1}{2}$ | 80 | | |

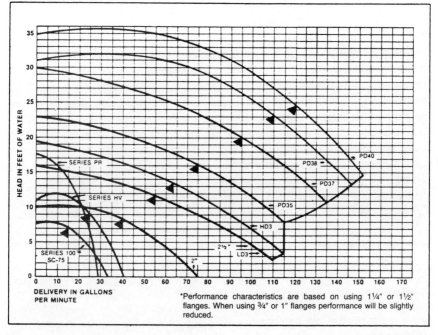

**Figure 7.4** Booster pump selection points.

Costs associated with booster pumps and series 60 in-line pumps for pipe sizes 2 in and less include

| | |
|---|---|
| Gate valve | 1 cap (on pipe support from floor) |
| Strainer | 8 nipples |
| Strainer blow-down valve | 2 reducers |
| Check valve | 1 gauge assembly including |
| Lever-operated plug valve | One $4\frac{1}{2}$-in dial pressure gauge |
| 1 elbow | Two $\frac{1}{4}$-in brass cocks |
| 1 tee | $\frac{1}{4}$-in piping to each side of pump |

No gauge assembly was included for series 100 or HV pumps.

Costs associated with booster pumps and series 60 in-line pumps for pipe sizes $2\frac{1}{2}$ and 3 in include

| | |
|---|---|
| Gate valve | 1 tee |
| Strainer | 1 weld cap (on pipe support from floor) |
| Strainer blow-down valve | |
| Silent check valve, wafer type | 1 flanged reducer |
| Lever-operated plug valve | 2 flanges, welded |
| 1 elbow | 2 companion flanges |
| | 1 gauge assembly |

Costs associated with end-suction base-mounted pumps include

Companion flanges (2)

Gate valve

Suction diffuser and screen

Diffuser drain valve and 5 ft of piping

Molded Teflon flexible pump connectors

$4\frac{1}{2}$-in dial pressure gauge assembly with two cocks and piping

Flanged reducer

Silent check valve, wafer type

Lever-operated plug valve

Inertia base on isolation springs

4-in concrete pad

B&G series 80 in-line pumps were priced with the following associated costs:

2 companion flanges

Gate valve

Strainer

Strainer blow-down valve

Molded Teflon flexible pump connectors

$4\frac{1}{2}$-in dial pressure gauge assembly with two cocks and piping

2 base elbows

Silent check valve, wafer type

Lever-operated plug valve

Inertia base on isolation springs

4-in concrete pad

Some models required the addition of reducers or increasers to the associated pipe; others did not. They were included in the pricing where appropriate.

Costs associated with double-suction base-mounted pumps include

2 companion flanges

Gate valve

Strainer

Strainer blow-down valve

Molded Teflon flexible pump connectors

$4\frac{1}{2}$-in dial pressure gauge assembly with two cocks and piping

2 base elbows, reducing where required

Silent check valve, wafer type

Lever-operated plug valve

Inertia base on isolation springs

4-in concrete pad

**Pot feeder assemblies**

Pot feeders were priced with $\frac{3}{4}$-in piping, two ball valves for shutoff, a $\frac{3}{4}$-in gate drain valve, and 10 ft of piping to drain. Unions and a $1\frac{1}{4}$-in air vent cock were included. The installed cost for the 2-gal variety was $523. The installed cost for the 5-gal type was $545.

## Antifreeze solution feeder assemblies

Two models were priced. Each included a 50-gal tank with lid; a $\frac{3}{4}$-in positive-displacement pump; low-water cutoff, alarm, and light; pressure gauge; and prewired control cabinet. The less expensive model is field-installed with both the tank and the pump mounted on a 4-in concrete pad; the installation is field-piped. The more expensive model is factory-mounted and -piped, with the tank on steel legs and the pump mounted on a bracket underneath; the 4-in pad is field-placed. This unit has a pressure-relief valve that is factory-piped back to the tank. The cost of field-erected assembly is $1,450; the cost of factory-mounted and -piped assembly is $1,800.

## Heating piping mains

Different sources show different prices for piping, with labor and material varying so that there seems to be no pattern. Some of the differences are attributable to the estimators pricing different components without a common base. Prices shown here in Tables 7.1a, b, and c are for threaded and coupled schedule 40 black steel pipe, for $\frac{3}{4}$- through 2-in sizes. For pipe sizes $2\frac{1}{2}$ through 8 in we used welded and flanged construction. Fittings are screwed 150-lb malleable iron for sizes through 2 in and 150-lb butt-weld fittings above 2 in. Flanges are 150-lb weld-neck. Strainers are 250-lb cast iron, screwed for sizes through 2 in and 125-lb cast iron flanged above 2 in. Valves are 150-lb brass or bronze for sizes through 2 in and 125-lb flanged cast iron above

TABLE 7.1a    Piping Costs: ASTM A-120 and A-53 Steel Pipe

| Pipe size, in | Horizontal piping, $/lin ft | | | | Vertical piping, $/lin ft | |
|---|---|---|---|---|---|---|
| | Pipe with no fittings; hangers for noninsulated pipe | Pipe with no fittings; hangers for insulated pipe | Pipe with tees each 30 ft; hangers for insulated pipe | Pipe with tees each 10 ft; hangers for insulated pipe | Pipe with no fittings; supports each floor | Pipe with tees each 12 ft; supports each floor |
| $\frac{3}{4}$ | 5.70 | 7.22 | 8.51 | 10.64 | 3.76 | 6.43 |
| 1 | 6.47 | 8.20 | 9.80 | 12.47 | 4.38 | 7.71 |
| $1\frac{1}{4}$ | 7.37 | 9.25 | 11.14 | 14.27 | 5.40 | 9.32 |
| $1\frac{1}{2}$ | 7.83 | 9.62 | 11.70 | 15.17 | 5.74 | 10.07 |
| 2 | 9.63 | 12.39 | 14.95 | 19.22 | 7.18 | 12.51 |
| $2\frac{1}{2}$ | 12.97 | 15.00 | 22.63 | 35.50 | 10.99 | 27.07 |
| 3 | 14.81 | 16.81 | 24.92 | 38.59 | 12.86 | 29.94 |
| 4 | 20.78 | 22.70 | 32.94 | 50.14 | 16.70 | 38.20 |
| 5 | 28.28 | 31.27 | 43.77 | 64.64 | 24.59 | 50.67 |
| 6 | 31.97 | 34.83 | 49.18 | 73.05 | 27.61 | 57.44 |
| 8 | 39.73 | 44.47 | 64.38 | 97.38 | 36.27 | 77.52 |

TABLE 7.1*b*  Piping Costs: Fittings and Miscellaneous Piping Accessories, $/each installed

| Steel pipe size, in | 90° elbows | Tees | Unions | Reducers | Weld-o-lets, thread-o-lets | Nipples, flanged, spools | Strainers, add blow-down valve | Teflon pump connectors |
|---|---|---|---|---|---|---|---|---|
| $\frac{3}{4}$ | 22 | 32 | 24 | — | 40 | 4 | 46 | — |
| 1 | 27 | 40 | 29 | 33 | 43 | 5 | 50 | — |
| $1\frac{1}{4}$ | 32 | 47 | 33 | 37 | 48 | 5 | 57 | — |
| $1\frac{1}{2}$ | 35 | 52 | 37 | 40 | 48 | 5 | 65 | 141 |
| 2 | 43 | 64 | 51 | 45 | 51 | 6 | 92 | 154 |
| $2\frac{1}{2}$ | 120 | 193 | — | 110 | — | 10/35 | 120 | 179 |
| 3 | 128 | 205 | — | 117 | — | 55 | 137 | 205 |
| 4 | 164 | 258 | — | 149 | — | 87 | 252 | 271 |
| 5 | 207 | 313 | — | 176 | — | 116 | 366 | 345 |
| 6 | 242 | 358 | — | 194 | — | 120 | 437 | 383 |
| 8 | 341 | 495 | — | 238 | — | 138 | 689 | 450 |

2 in. Plug valves are 175-lb semisteel. Both the material cost and the total cost of the flanges are shown. Use the material cost only when welding to any fitting such as a reducer, tee, or elbow. The welding cost is included in these fittings. These tables may be used when the estimator needs to do a piping takeoff type of estimate.

In investigating the costs of piping, hangers, and fittings it became apparent that no single cost per foot would be applicable to every situation. We eventually arrived at six sets of prices which might approximate steel piping costs for estimating purposes. Because the hanger costs materially affect the price per foot of piping, two basic sets of prices evolved: one for horizontal piping run overhead and the other for vertical piping, with assumed 12-ft floor-to-floor heights. We then have several variations on the horizontal and vertical piping.

For horizontal piping we have costs for steel pipe with no fittings except for coupling or welded joints and with hangers for noninsulated pipe. This might be used for condenser piping and other noninsulated systems. We also have costs for insulated steel pipe with hangers but with no fittings except couplings or welded joints. To this cost we have added two variations for conceptual estimating. One for perimeter heating mains which might be used for classroom unit ventilators or other similar systems with reducing tees every 30 ft and an elbow every 100 ft, and a second for perimeter mains for fan coils or other similar systems with reducing tees every 10 ft and an elbow every 100 ft. For these systems, we used an Auto-Grip clevis-type hanger with threaded rod. For insulated pipes we added a sheet metal pipe shield and sized the hanger and shield for insulation. We used a beam C clamp price for piping up through 3 in and a welded beam attachment

**TABLE 7.1c  Piping Costs: Valves and Miscellaneous Piping Accessories, $/each installed**

| Steel pipe size, in | Flanges, mat'l/ total | Gate valve | Swing check | Silent wafer check | Globe | Eccentric non-lubricated plug | Ball | Butterfly | Balancing circuit setter | Two-way temperature control valve* | Three-way temperature control valve* |
|---|---|---|---|---|---|---|---|---|---|---|---|
| $\frac{3}{4}$ | — | 30 | 34 | — | 41 | 47 | 28 | — | 66 | 44 | 58 |
| 1 | — | 38 | 43 | — | 57 | 55 | 34 | — | 79 | 52 | 67 |
| $1\frac{1}{4}$ | — | 48 | 54 | — | 76 | 71 | 44 | — | 99 | 58 | 76 |
| $1\frac{1}{2}$ | — | 54 | 60 | — | 89 | 82 | 51 | — | 112 | 65 | 83 |
| 2 | — | 68 | 82 | 72 | 126 | 96 | 63 | — | 149 | 74 | 99 |
| $2\frac{1}{2}$ | 20/92 | 172 | 116 | 74 | 221 | 150 | — | 95 | 264 | 91 | 129 |
| 3 | 23/98 | 200 | 129 | 83 | 250 | 179 | — | 101 | 354 | 91 | 158 |
| 4 | 30/138 | 295 | 214 | 107 | 409 | 292 | — | 143 | 484 | 169 | 276 |
| 5 | 38/159 | 445 | 308 | 140 | 621 | 499 | — | 181 | — | 209 | 303 |
| 6 | 45/174 | 457 | 334 | 187 | 636 | 499 | — | 204 | — | 246 | 309 |
| 8 | 73/232 | 709 | 659 | 326 | 1223 | 674 | — | 266 | — | 269 | 334 |

*Installation only, see Temperature Control section for material cost.

for sizes 4 through 8 in. Hanger labor and material should be valid for inserts for plywood decks, self-drilling shields, and concrete deck inserts. For Q deck, plastic, or steel pan deck add $1.00/ft to these costs.

For vertical piping the hangers used were beam clamps at every floor which are good for insulated as well as noninsulated piping. The first set of costs is for piping with no fittings, the second set is for piping with a reducing tee every 12 ft (at each floor level).

Labor is based on pipe fitters doing the work for pipe up through 2 in and for labor helpers doing 33% of the work for pipe $2\frac{1}{2}$ through 8 in.

### Perimeter heating systems

**Distribution mains.**    The curves for Figure 7.5 show the cost of mains in dollars per MBH for floor areas which the mains serve. All are for radiation zones of approximately 30 ft in length.

The cost of mains is one of the more difficult amounts to ascertain before any design work is done because the piping can be routed in many different fashions. If your system is very nearly like the one used

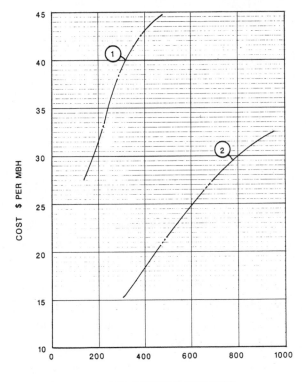

MBH PERIMETER HEAT LOSS

**Figure 7.5** Perimeter heating systems, distribution mains. (1) Mains serving one floor; (2) mains serving two floors.

here, these costs are accurate; if the system is different, these costs should only be used in the early stages before any design work is done and should be treated more as allowances than as actual costs. Later, when preliminary design is done, a piping takeoff cost may be completed using Tables 7.1a, b, and c. These costs are based on wall losses plus infiltration at the perimeter of 478 Btu per lineal foot (see discussion on radiation) and a rather square-shaped building (or building wing) with the boiler room or heating source in one corner of the building. Mains are assumed to be overhead in the ceiling space and systems are always reverse-return with no main smaller than 1 in. The curves show costs of piping serving one story with no core loss or piping for same. For piping serving two stories, the perimeter piping serves both up and down from the intermediate ceiling, again with no core loss or piping for same. Using Tables 7.1a and b, add any transmission piping costs from the heating plant to the point where distribution mains start, including mains from floor to floor where there are more than two floors. Boiler room piping costs are included with the boiler room, Figure 7.1. Runouts, fittings, valves, etc., at the terminal units are included with those costs.

Costs in dollars per MBH are based on a $20°dt$; for a $40°dt$ system the piping can handle twice the MBH shown. So the cost in dollars per MBH for 500 MBH is the same as for 1,000 MBH at $40°dt$. Also see the discussion later in this chapter of end-fed vs. center-fed systems under Perimeter Piping Mains for Fan Coils and Unit Ventilators.

### Primary-secondary piping bridle assemblies

Primary-secondary piping is often employed in building complexes today. We have priced the bridle that is required at each secondary pumping loop. Table 7.2 shows these costs, which include a venturi, gate valve, eccentric nonlubricated plug valve, two ells, and two tees for secondary piping connections. Two-inch and smaller sizes are screwed and unions are included; $2\frac{1}{2}$ in and larger sizes are welded and flanged.

**TABLE 7.2    Primary-Secondary Piping Bridle Assemblies**

| Pipe size, in | Flow, gpm | Total cost each assembly, $ |
|---|---|---|
| $1\frac{1}{2}$ | 25 | 542 |
| 2 | 50 | 728 |
| $2\frac{1}{2}$ | 80 | 1,542 |
| 3 | 140 | 1,765 |
| 4 | 300 | 2,344 |
| 5 | 500 | 3,067 |

NOTE: For each meter required, 6-in dial, add $800.

TABLE 7.3  Small Reheat Coil Assemblies

| Coil size | cfm | MBH | Total $ | $/cfm | $/MBH |
|---|---|---|---|---|---|
| | | One-Row Coils | | | |
| 12 × 12 | 700 | 30.6 | 755 | 1.08 | 24.67 |
| 12 × 18 | 1,050 | 46.2 | 864 | .823 | 18.70 |
| 12 × 24 | 1,400 | 61.6 | 872 | .623 | 14.16 |
| 18 × 24 | 2,100 | 92.4 | 955 | .474 | 10.77 |
| 18 × 30 | 2,625 | 115.5 | 1,005 | .383 | 8.70 |
| | | Two-Row Coils | | | |
| 12 × 12 | 600 | 39.6 | 782 | 1.30 | 19.75 |
| 12 × 18 | 900 | 59.4 | 897 | .997 | 15.10 |
| 12 × 24 | 1,200 | 79.2 | 910 | .758 | 11.49 |
| 18 × 24 | 1,800 | 118.8 | 1,051 | .584 | 8.85 |
| 18 × 30 | 2,250 | 148.5 | 1,071 | .476 | 7.21 |

**Small hot-water reheat coil assemblies**

Five sizes of coils were priced, from 1 to 3.75 ft$^2$ of face area, using Trane Company style ST coils. One-row coils were selected at 700-ft/min face velocity, and two-row coils were selected at 600-ft/min face velocity to give a pressure loss of about .2-in wg.

Our assumption on the two-row coils was to heat the air quantity used for cooling approximately 60°F [55° entering air temperature (EAT), 115° leaving air temperature (LAT)]. This assumption was based on 1.4 cfm/ft$^2$ of floor area required as a minimum for cooling with a wall loss of 480 Btu per lineal foot and a roof loss of 8 Btu/ft$^2$. For 1 lineal foot of wall and the space 10 ft deep, this would amount to 1.4 cfm at about a 40° rise above room temperature to cover the loss plus a 20° rise from 55° LAT to 75° room temperature. The one-row coils would be suitable for spaces with roof loss, interior rooms, and perimeter spaces with small losses.

Table 7.3 shows installed costs in dollars per cubic foot per minute and in dollars per MBH. Costs include piping with 15-ft piping runouts, six elbows, two unions, a gate valve, balancing valve, air vent, and the installation of a two-way temperature control valve.

**Hot-water unit heater and cabinet heater assemblies**

Heaters were picked at 200° entering water temperature (EWT), 20°$dt$, and 60° EAT. Prices are from the Trane Company, La Crosse, Wisconsin.

Unit heaters priced were type S, horizontally mounted, and type P, vertically mounted with a louver cone diffuser. Heaters were picked at the higher rotation speed cataloged.

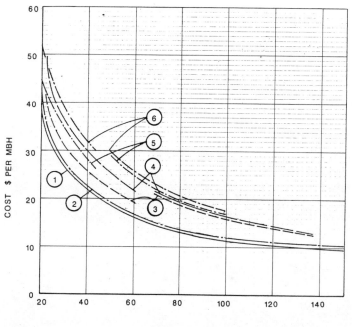

MBH CAPACITY

**Figure 7.6**   Unit heater and cabinet heater assemblies. (1) Model S unit heater; (2) model P unit heater; (3) model B cabinet heater, floor-mounted; (4) model D cabinet heater, ceiling-mounted; (5) model B heater, slow-speed; (6) model D heater, slow-speed.

Cabinet heaters priced were type B, floor-mounted, and type D, ceiling-mounted with grilled back and gray baked-enamel finish. All were picked at the high rotation speed. One-inch glass fiber throwaway filters were included. A cost curve (Figure 7.6) is also shown if the units were picked at the lower speed for quietness.

All prices are installed costs and include piping with 15-ft piping runouts, six elbows, two unions, gate valve, balancing valve, six nipples, and automatic air vent (see Figure 7.6).

### Hot-water convector assemblies

Curves in Figure 7.7 show the installed cost of three styles of convectors; each style was plotted with two depths, 6 and 8 in. Prices were based on Trane Company convectors: model W, wall-hung; model FG, floor-mounted with front inlet grille and front outlet grille; and model SG, semirecessed floor-mounted with front inlet grille and front outlet grille and gray baked-enamel finish. All units were picked at 200° EWT, 20°$dt$, 65° EAT. Capacity correction factors because of grilles were in the range of 2 to 5% and were ignored.

Fully recessed models cost less than the semirecessed models. Their

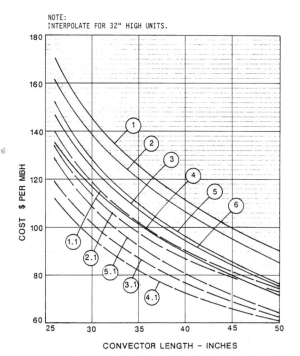

NOTE:
INTERPOLATE FOR 32" HIGH UNITS.

CONVECTOR LENGTH – INCHES

**Figure 7.7** Convector assemblies. (1) Model SG, 26 in high, 6 in deep; (2) model SG, 38 in high, 6 in deep; (3) model FG, 26 in high, 6 in deep; (4) model FG, 38 in high, 6 in deep; (5) model W, 26 in high, 6 in deep; (6) model W, 32 in high, 6 in deep. (1.1) Model SG, 26 in high, 8 in deep; (2.1) model SG, 38 in high, 8 in deep; (3.1) model FG, 26 in high, 8 in deep; (4.1) model FG, 38 in high, 8 in deep; (5.1) model W, 26 in high, 8 in deep.

installed costs are approximately 88% of the cost of semirecessed models.

Installed costs include 15-ft piping runouts, eight elbows, two unions, gate valve, balancing valve, nine nipples, auto air vent, and installation of a control valve (see Figure 7.7).

**Radiation**

**Trial heat losses of buildings.** Trial heat losses are calculated for a 12-ft-high wall that is 10 ft long with $80°dt$ in a building from the old radiator days (which wall fin radiation replaced). It has the following characteristics:

No ceiling space

Brick or stone wall 12-in thick with plaster inside

Window 4 ft wide, 6 ft high, double-hung, wood

Wall U factor = .53
Window U factor = 1.13
Infiltration = 39 cfh per lineal foot from a crack around window (ignore infiltration through wall)
Wall loss = $(120 \text{ ft}^2 - 24 \text{ ft}^2 \text{ window}) \times .53 \times 80°dt = 4{,}070$ Btu/hr
Window loss = $24 \text{ ft}^2 \times 1.13 \times 80°dt = 2{,}170$ Btu/hr

$$\text{Infiltration} = \frac{39 \text{ cfh/lineal foot} \times 24\text{-ft crack}}{60 \text{ min/hr}} \times 1.1 \times 80°dt = 1{,}373 \text{ Btu/hr}$$

Skin heat loss (10-ft-long wall) = 7,613 Btu/hr.
Sub-total loss per lineal foot = 761 Btu

If the roof loss were added for a 10-ft-wide room with a flat concrete roof without insulation:

U = .4
$100 \text{ ft}^2 \times .4 \times 80°dt = 3{,}200$ Btu/hr
Room heat loss, 10-ft-wide room = 10,813 Btu/hr
Total loss per lineal foot = 1,081 Btu

Losses for today's so-called all-glass building (a worst case) are calculated as follows:

Ceiling height = 9 ft
Insulated metal wall panels, U = .12
Double glazing, $\frac{1}{2}$-in air space, U = .55
Infiltration = one-half air change per hour
Panel loss = 4 ft high $\times$ 10 ft $\times$ 12 $\times$ 80°dt = 384 Btu/hr
Glass loss = 8 ft high $\times$ 10 ft $\times$ 55 $\times$ 80°dt = 3,580 Btu/hr
Infiltration = 10 ft wide $\times$ 10-ft-long room $\times$ 12 ft high

$$= \frac{1200 \text{ ft}^3 \times \frac{1}{2} \text{ air change/hr}}{60 \text{ min/hr}} \times 1.1 \times 80°dt = 880 \text{ Btu/hr}$$

Skin heat loss (10-ft-long wall) = 4,784 Btu/hr
Subtotal loss per lineal foot = 478 Btu

A roof of today has a U value of .08 to .1 for a room 10 ft wide:

$100 \text{ ft}^2 \times .1 \times 80°dt = 800$ Btu/hr
Room heat loss, 10-ft-wide room = 5,584 Btu/hr
Total loss per lineal foot = 558 Btu

**Discussion.**    Finned-tube radiation replaced radiators years ago. Its greatest benefit was in uniform heating along the entire perimeter rather than having a hot spot at the radiator and then a long cold wall.

A look at the trial heat losses we completed, along with studying most radiation catalogs, shows us that the radiation capacity is generally too large to match today's building losses. Even a building with extreme winter temperatures, say $120°dt$ from outdoor to indoor temperature, would only need radiation with a capacity of

$$558 \text{ Btu/lineal foot} \times \frac{120°dt}{80°dt} = 837 \text{ Btu/lineal foot}$$

Of course there are applications for high-output radiation. Any two-story building space, renovation of older building, etc., might require higher-output radiation, but in studying the radiation catalogs, one-row radiation with an enclosure only 14 in high has capacities up to 1,500 Btu/lineal foot. So if we are to follow the doctrine of providing comfort in our buildings, it would seem that we should look for and specify radiation with a far lower output than we have in the past. This output should match the heat loss as closely as possible.

**Conclusion.**    Building skin losses are averaging 500 Btu per lineal foot or less. Three-quarter-inch piping will handle 40,000 Btu/hr or 80 ft of radiation (if it could be equated to loss) per zone. Studies of our past projects show wall fin radiation with an average radiation zone of about 30 ft. Three-quarter-inch pipe can handle this load, and $\frac{3}{4}$-in tube gives the best buy in dollars per Btu for radiation. The larger the tube, the higher the cost and the less the Btu per dollar value. For example, $\frac{3}{4}$-and 1-in tubes are cost-efficient, but $1\frac{1}{4}$-in and larger tubes seem less cost efficient, possibly because they are large enough to restrict air flow; therefore, there is less output and more dollar cost.

From the studies we have made, the best buy for a radiation element in dollars per Btu is the smallest element available for any given enclosure. This is the element with the smallest tube and with the smallest fin size; however, 40 to 50 fins per foot seem more cost efficient than 32 fins per foot.

Steel elements are so much more expensive than copper or aluminum that it is hard to see why one would specify them. Two comparable elements in Btu output for radiation cost are as follows:

Cu/Al: 640 Btu/ft − $2.92 lineal foot = $ 4.56/MBH
Steel: 650 Btu/ft − $7.73 lineal foot = $11.89/MBH

**Estimating costs.**    Using the Sterling Radiation catalog, two models which might be suitable for use in today's commercial and institutional buildings are (1) the classic Slimline with the smallest fin tube ($\frac{3}{4}$ in, $2\frac{1}{4}$ in $\times$ $2\frac{1}{2}$ in, 50 fins per foot), which has a capacity of 640 Btu/ft at 190°F average water temperature and an $11\frac{7}{8}$-in-high enclosure. This is a good selection for any application where the best looking radiation is desired. Higher outputs can be obtained when needed by going to elements with $\frac{3}{4}$- or 1-in tubes and capacities up to 900

TABLE 7.4   Costs of Radiation Assemblies

(Based on Sterling Radiator, Westfield, Massachusetts)

| Style | Capacity, Btu/ft | Total cost, $ | Cost per foot, $ | Cost per MBH, $ |
|---|---|---|---|---|
| Commercial and Institutional Projects, 30-ft Assemblies* | | | | |
| Classic VA-AR-14 | 1,010 | 1,953 | 65.10 | 74.37 |
| Slimline VK-AR-14 | 640 | 1,765 | 58.83 | 106.07 |
| Slimline VK-AR-14 | 820 | 1,769 | 58.97 | 82.97 |
| Versa-Line JVA-14: | | | | |
| T Encl | 1,030 | 1,583 | 52.77 | 59.11 |
| S Encl | 1,150 | 1,574 | 52.47 | 52.64 |
| LB-2 | 790 | 1,350 | 45.00 | 65.72 |
| Hi-Pak | 950 | 1,300 | 43.33 | 52.63 |
| Senior | 830 | 851 | 53.19 | 76.92 |
| Kom-Pak | 650 | 800 | 50.00 | 92.33 |

\* Thirty-foot assemblies have three 9-ft lengths of element and approximately 26 ft of finned tube. Sixteen-ft assemblies have two 7-ft lengths of element and approximately 13.3 ft of finned tube.

Btu/lineal foot. (2) Sterling LB-2 commercial and institutional base-board radiation is also a reasonable choice for outputs in the 700- to 900-Btu/lineal foot range. It has the traditional sloping top, stamped grille look, and might be used where this appearance is acceptable or preferred.

Installed costs are shown for several different styles (Table 7.4). These are based on 30-ft zones for commercial and institutional projects and 16-ft zones for residential projects. Costs include a baked-enamel enclosure with end and corner caps and two valve-access compartments. Elements are the smallest available for that enclosure and are copper and aluminum, picked at 200° EWT, 20°$dt$. The piping includes 15-ft piping runouts, five elbows, one reducing tee and bushing, one air vent, two dielectric unions, one gate valve and balancing valve, plus the installation of a control valve and soldered copper connections between element lengths.

Dampers are not included except on the Sterling Hi-Pak, Senior, and Kom-Pak models. Dampers for other models run $2.40/ft. No back panel is included on the Versa-Line models because it is not standard.

Since it would be hard to match the heat loss exactly with the number of feet of radiation heater selected, some efficiency factor should be applied. To determine the cost, it is first necessary to decide what percentage of the heat loss is to be handled by the radiation. Take that number and divide by the Btu's per hour output of the radiation heater selected to get the footage required. Multiply that number by 1.2 to get the estimated radiation footage for the project. Multiply the estimated radiation footage by the theoretical cost per foot, shown in Table 7.4, to get the radiation cost for the project.

Costs in dollars per MBH are shown in Table 7.4 for comparison to other heating methods. These costs are based on picking radiation that will fill up 75 to 85% of the wall space with fin tube to cover the heat losses with the enclosure running completely from wall to wall. If the radiation has more capacity than this and requires a great deal more enclosure than fin tube, then the values are not accurate.

### Radiant ceiling hot-water panel assemblies

The graph in Figure 7.8 shows the installed cost for Airtex 2 × 4 ft radiant panels. Airtex makes two models: type H, which is used for heating, and type HPH (high-performance heating), which can also be used for cooling. Also available are extruded aluminum panels in various widths and lengths.

Panels were picked at 200° EWT, 20°$dt$, and 70° room air temperature. At these conditions, we have the following capacities:

| | |
|---|---|
| H panel | 2 × 4 ft 1,600 Btu/hr |
| HPH panel | 2 × 4 ft 1,820 Btu/hr |
| Extruded | 20 in wide = 417 Btu/ft = 1,668 Btu/hr for 4 ft |

Studies of our past projects show that wall fin radiation has been zoned with an average zone of about 30 ft. Equating a radiant panel

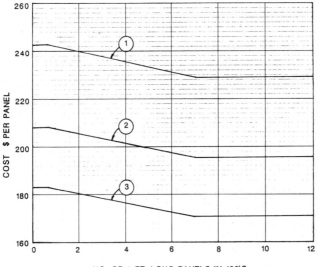

NO. OF 4 FT. LONG PANELS IN 100'S

**Figure 7.8**   Radiant ceiling hot water panel assemblies: cost per panel. (1) Twenty-inch-wide extruded aluminum panels; (2) HPH panels; (3) H panels.

zone to a radiation zone, 30 ft of wall per heating zone would be approximately eight panels at 400 to 455 Btu per lineal foot. This capacity will handle the loss at 0° outside air temperature from double glazing that is about 8 ft high. Eight panel zones also give a reasonable pressure loss for the pump. We have, therefore, priced these panels on an eight-panel-per-piping-zone basis. The piping includes 15-ft piping runouts, six elbows, two dielectric unions, seven nipples, one gate valve and balancing valve, plus the installation of a control valve and soldered copper connections between all panels. Cost includes installation of the panels in a ceiling system which has been installed by others.

To equate the extruded aluminum cost to the 2 × 4 ft panel cost, we used three 10-ft lengths of panel. The total price of the 30-ft length divided by the Btu per hour output gives a dollars per MBH figure. To compare the cost of 2 × 4 ft panels to the 20-in-wide extruded panels, take the total cost of 30 ft of the extruded assembly and divide by 30 ft; then multiply by 4 ft. This gives the theoretical cost of a 4-ft panel.

Since it would be hard to match the heat loss exactly with the number of panels selected, some efficiency factor should be applied when selecting the number of panels required. It is first necessary to determine what percentage of the heat loss is to be handled by the radiant panels. Take that number and divide by the Btu per hour output of the panel system selected to get the theoretical number of panels. Multiply the theoretical number by 1.10 to get the estimated number of panels for the project. Enter the chart for this estimated number of panels and read the cost in dollars per panel.

A separate graph (Figure 7.9) was drawn and is included to show the cost in dollars per MBH. It may be used to compare the dollar value with other terminal heating methods.

### Electric heaters

Electric unit heaters and cabinet heaters were priced along with electric radiation, electric radiant ceiling panels, and several specialty heaters that we seem to find a need for from time to time (see Exhibit 7.2). All prices are from Chromalox, Emerson Environmental Products, a division of Emerson Electric Company, Hazelwood, Missouri. Prices are with standard accessories as shown in their catalog except as may be noted here.

**Unit heaters.**   Electric unit heaters were priced using the Chromalox MUH series, either horizontally or vertically mounted. For vertically mounted units, louver face diffusers were added to the cost. Note that if the voltage is 208 V, the units must be specified for 208 V rather than the combination 208/240 V. The capacity of a combination voltage unit at 208 V is approximately 75% of the unit's capacity at 240 V,

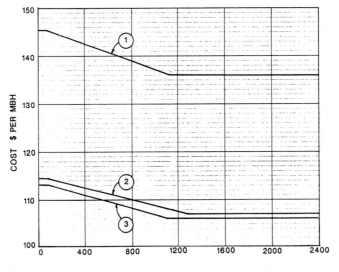

MBH

**Figure 7.9**  Radiant ceiling hot water panel assemblies: cost per MBH. (1) Twenty-inch-wide extruded aluminum panels; (2) HPH panels; (3) H panels.

**Exhibit 7.2**
**ELECTRIC HEATING COST ESTIMATING SUMMARY SHEET**

|  | Total cost, $ |
|---|---|
| Electric unit heaters: | |
|   Horizontal mounting | _____ |
|   Vertical mounting | _____ |
| Electric cabinet heaters: | |
|   Floor-mounted | _____ |
|   Ceiling-mounted | _____ |
| Electric radiation: | |
|   Residential style | _____ |
|   Commercial-institutional: | |
|     Style CSH | _____ |
|     Style DSH | _____ |
| Electric radiant ceiling panels | _____ |
| Specialty heaters: | |
|   Kickspace heaters | _____ |
|   Floor heaters | _____ |
|   Wall heaters | _____ |
| Miscellaneous_____ | _____ |
| Electric heating portion estimate, this sheet | _____ |

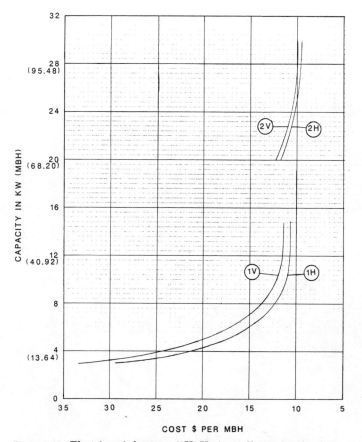

**Figure 7.10**   Electric unit heaters. (1H) Horizontally mounted, capacities up through 15 kW; (2H) horizontally mounted, capacities of 20 to 30 kW; (1V) vertically mounted, capacities up through 15 kw; (2V) vertically mounted, capacities of 20 to 30 kW.

whereas the 208-V unit has 100% of the 240-V unit capacity. The units have integral 24-V control thermostats. The cost per MBH for these units decreases as the unit size increases up through 15 kW. The cost per MBH increases for the 20-kW unit and then diminishes through 30 kW. The curves in Figure 7.10 reflect this. Costs are installed costs, but with no disconnect or power wiring included.

**Cabinet heaters.** Electric cabinet heaters were priced using Chromalox CUF or CUI (floor-mounted) and CUC (ceiling-mounted) units with 24-V controls, a wall-mounted 24-V thermostat, baked-enamel finish, and throwaway filters. Units are 208, 240, or 277 V. Each cabinet heater length comes with a choice of two, three, four, five, or six elements for increased output. The cost per MBH decreases with

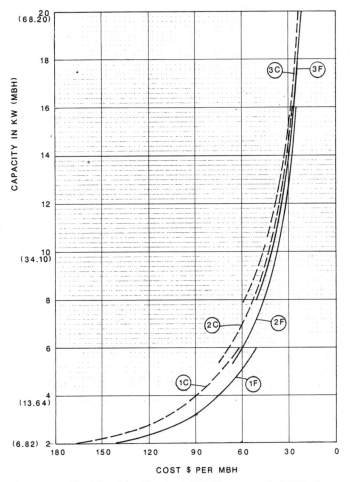

**Figure 7.11**  Electric cabinet heaters. (1F) Floor-mounted, 36 in long; (2F) floor-mounted, 58 in long; (3F) floor-mounted, 80 in long; (1C) ceiling-mounted, 36 in long; (2C) ceiling-mounted, 58 in long; (3C) ceiling-mounted, 80 in long.

the number of elements. Cabinet lengths of 36, 47, 58, 69, and 80 in were priced. Because the curves were so close together, only the cost curves for 36, 58, and 80 in, plotted for output with two, three, four, five, and six elements, are shown (see Figure 7.11). The shorter units are always cheaper per MBH when the capacities overlap with the longer units. Costs are installed costs, but with no disconnect or power wiring included.

**Baseboard radiation.**    Zones for hot-water radiation were set up on 16 ft of radiation for residential zones and 30 ft of radiation for commercial

TABLE 7.5    Electric Radiation Costs

| Radiation* | Watts/foot | Zone size | Total cost, $ | Cost, $/kW | Cost, $/ MBH |
|---|---|---|---|---|---|
| | | Residential | | | |
| BBF—8-ft sections† | 187.5 | 16 | 484 | 161.33 | $ 51.55 |
| BBF—8-ft sections‡ | 250 | 16 | 484 | 121.00 | 38.66 |
| BBF—4-ft sections‡ | 187.5 | 16 | 505 | 168.33 | 55.55 |
| BBF—4-ft sections‡ | 250 | 16 | 505 | 126.25 | 41.66 |
| | | Commercial-Institutional§ | | | |
| CSH-07—10-ft sections | 125 | 30 | 1,712 | 456.53 | 137.94 |
| 10-ft sections | 250 | 30 | 1,912 | 254.93 | 77.72 |
| DSH-07—10-ft sections | 125 | 30 | 1,838 | 490.13 | 147.78 |
| 10-ft sections | 250 | 30 | 2,055 | 274 | 83.30 |

* All models are Chromalox.
† Eight-foot-long sections, 208 or 240 V.
‡ Four-foot-long sections, 120 V.
§ All commercial units are 208, 240, or 277 V.
NOTE: For a 24-V wall thermostat and wiring, add $75.

and institutional zones. In order to have a common base for comparison, electric radiation was calculated on the same zone basis with one low-voltage wall thermostat per zone.

Residential radiation was priced using Chromalox BBF Rebel II radiation units, with either a 187.5-W/ft (640 Btu/ft) element or a 250-W/ft (853 Btu/ft) element. These are comparable outputs to the two hot-water elements picked for residential service.

Two styles of commercial-institutional radiation were priced: the Chromalox CSH (commercial sill height), CSH-07, with 7-in-high enclosure, and the Chromalox DSH (decorative sill height), DSH-07, with 7-in-high enclosure. Each was priced with two elements: 125 W/ft (426 Btu/ft) and 250 W/ft (853 Btu/ft).

All radiation units come with a baked-enamel cover and 1 24-V transformer relay. Commercial-institutional radiation units also have filler sections, control sections, end caps, and corners included.

The costs of a 24-V thermostat and wiring are shown as extra cost. Power source wiring and disconnect are not included in the total cost or cost per kilowatt. Use this cost for estimating (see Table 7.5).

To arrive at the cost per MBH we have added the power wiring to connect the sections together. Use this price to compare with the cost of hot-water radiation.

**Radiant ceiling panels.** Radiation zones for hot-water radiant panels were set up on eight panel zones (see Figures 7.8 and 7.9 and text discussion). Panel capacities for 2 × 4 ft panels ranged from 1,600 to

1,820 Btu per panel; 2 × 4 ft electric panels rated at 500 W (1,707 Btu) would be comparable.

Panels priced were Chromalox style CP for T-bar ceiling mounting, 208, 240, or 277 V, eight panels per zone, with a 24-V transformer-relay and all connecting power wiring.

The costs of a 24-V thermostat and wiring are shown as extra cost. Power source wiring and disconnect are not included in the total cost or cost per kilowatt. Use this price for estimating.

To arrive at the cost per MBH we have added the power wiring to connect the eight panels together. This amounts to three junction boxes and eight flex connections to the panels. Use this price to compare with the cost of hot-water panels:

| Panels | Watts each | Zone size | Total cost, $ | Cost, $/kW | Cost, $/MBH |
|--------|-----------|-----------|---------------|------------|-------------|
| Ceiling | 500 | 8 panels | 812 | 204 | $70.59 |

For a 24-V wall thermostat and wiring, add $75.

**Kickspace heater.** For a kickspace heater, we used the Chromalox KSH, rated at 450 W, 110 V, with an integral line voltage thermostat. Installed, but with no disconnect or power wiring, the cost was $250 each.

**Floor heater.** For a floor drop-in heater, we used the Chromalox FDI, rated at 375 to 1500 W, 110 V, with an integral line voltage thermostat. Installed, but with no disconnect or power wiring, the cost was $270 each.

**Wall heater.** For a wall heater, we selected the Chromalox AWH-4000, rated 1,500/2,000 W and 3,000/4,000 W, 208 or 240 V, with an integral thermostat. Installed, but with no power wiring, the 1,500/2,000-W unit was $360 each and the 3,000/4,000-W unit was $380 each.

### Cost comparison curves for heating terminal units

To give an approximation of the comparative costs of different terminal heating units, we have plotted the curves on Figure 7.12. The following list keys the curves to their corresponding heating terminal units:

1. Model S unit heater, horizontally mounted; least expensive unit heater

2. Model B cabinet heater, floor-mounted; least expensive cabinet heater

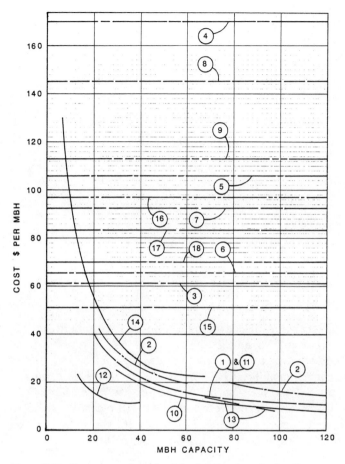

**Figure 7.12**   Cost comparison chart for heating terminal units.

3. Model FG convector, 38 in high × 50 in long × 8 in deep; least expensive convector

4. Model SG convector, 26 in high × 26 in long × 6 in deep; most expensive convector

5. Architectural radiation, Slimline, 640 Btu/ft; most expensive radiation

6. Commercial radiation, LB-2, 790 Btu/ft

7. Residential radiation, Kom-Pak, 650 Btu/ft

8. Radiant ceiling panels, 20 in wide extruded; most expensive radiant panel

9. Radiant ceiling panels, H panel; least expensive radiant panel

10. One-row reheat coils

11. Two-row reheat coils (closely parallels curve 1)

12. Electric unit heaters, horizontally mounted to 15 kW; least expensive electric unit heater

13. Electric unit heaters, horizontally mounted, 20 to 30 kW

14. Electric cabinet heater, floor-mounted; least expensive electric cabinet heater

15. Electric residential radiation, 8-ft section, 640 Btu/ft

16. Electric architectural radiation, 10-ft sections, 853 Btu/ft

17. Electric commercial radiation, 10-ft sections, 853 Btu/ft

18. Electric radiant ceiling panels

## Cooling

### Chiller Plants

Four types of chillers were priced using Trane Co. equipment as the pricing source. Figures 7.13 and 7.14 show the costs of the various assemblies in dollars per ton. Curves are for hermetic *centrifugal* water-cooled chillers, model CVHE; hermetic *reciprocating* water-cooled chillers, model CGWB; hermetic reciprocating *condenserless* chillers, model CCA; and hermetic reciprocating *air-cooled* chillers, model CGA. Separate figures (Figures 7.15 and 7.16) show the costs of (1) air-cooled condensers, model CAU, for use with the reciprocating condenserless chiller, and (2) cooling towers to use with water-cooled chillers.

All reciprocating chillers were picked with dual compressors for protection against being completely down. Conditions were 43°F leaving water temperature (LWT), 10° rise, 2.4 gpm/ton. Water-cooled condensers were picked at 80° EWT, 10° rise. For climates where the wet bulb is 70° and above and 80° water off the tower would be difficult or impossible to generate, I have reviewed the capacities using 85° EWT and a 10° rise, and they appear to decrease slightly less than 3%. This correction factor could be used for climates having wet-bulb temperatures from 70° up to 75°. Air-cooled equipment was chosen at 105° condensing for 95° ambient. For 100° ambient and 110° condensing the capacities diminish by 4%.

Chiller costs include labor to install the unit, furnishing a starter with control transformer, under- and overcurrent protection, phase

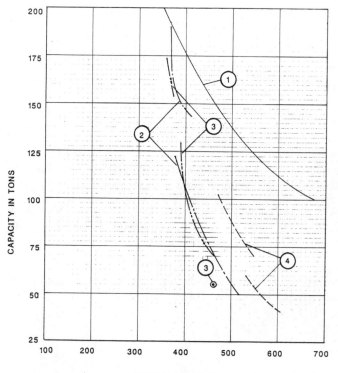

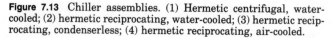

COST $ PER TON

**Figure 7.13** Chiller assemblies. (1) Hermetic centrifugal, water-cooled; (2) hermetic reciprocating, water-cooled; (3) hermetic reciprocating, condenserless; (4) hermetic reciprocating, air-cooled.

failure, and reversal protection and hand-off-automatic (HOA) switch. Where filter-driers were not standard, they were added, and where motors had power factors less than .95, power factor correction was added to increase the power factor to .95. For centrifugal machines we added the cost of factory insulation and voltmeter and ammeter. We estimated a 4-in concrete pad for mounting centrifugal machines and air-cooled chillers mounted outdoors. For reciprocating machines in the building we added the cost of an isolation base on springs.

The curves reflect the installed cost of the chilled water generating system except for pumping (see Figures 7.2 and 7.3), domestic water feeder assembly, and chemical feeder assembly or antifreeze feeder assembly. The condensers are included only on the packaged hermetic reciprocating air-cooled chillers. The air-cooled condensers and piping, and cooling towers, pumping, and piping are not included and have separate figures. Add these costs to the chiller plant cost. Included in the chiller plant pricing for each size of chiller studied are

Chiller and mounting base

Air separation system

Expansion tank and air fitting

Relief valve and piping to drain

Water feeder assembly

Chiller shutoff valves (two gate valves)

2 molded Teflon flexible connectors

2 spool pieces with temperature and pressure taps

2 elbows to chiller

## Chiller standards

**Isolation bases.**  These are steel bases with a reinforced concrete core on spring vibration isolators, from Mason Industries.

**Expansion tank.**  Tanks were chosen using average sizes from past projects of varying types with up to three stories of piping above the tank. ASME tanks with sight glass and air fitting were used.

**Air separation system.**  The B&G Rolairtrol air separation (without screen) system was priced. Included is the piping to the expansion tank and drain valve with 10 ft of drain piping.

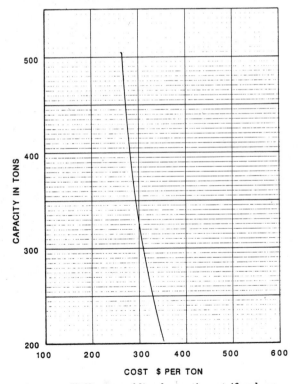

**Figure 7.14**  Chiller assemblies: hermetic centrifugal, water-cooled, 200 to 500 tons.

**Chiller room piping.** Piping in the equipment rooms on past projects was studied, and for chilled water piping lengths we added 120 ft of piping on the 50-ton sizes, graduating up to 250 ft at 500 tons.

### Water feeder assembly

The water feeder assembly consisted of a $\frac{3}{4}$-in vacuum breaker and B&G no. 12 water feeder in a three-valve bypass with a vacuum breaker drain pipe to drain. This is plumbing work, but it is included here for ease in pricing. The cost is $825.

### Air-cooled condenser assemblies

Figure 7.15 shows the cost of air-cooled condensers, the Trane Company CAUB and CAUA models. Costs include the unit or units necessary to meet the capacities of the chiller at 40°F suction and include the necessary piping between the condenser and the chiller. Chillers were chosen with dual compressors. The Trane Company recommends separate circuit piping for each compressor-condenser-chiller circuit, and piping was designed and priced that way. Included in the piping were bronze corrugated-metal flexible connectors at the compressor, chiller, and condenser, and shutoff valves at the condenser for each condensing circuit. A 75-ft distance between the condenser and chiller was assumed, and piping was designed with a double-riser system using ACR type L copper pipe with 95-5 solder. Piping was priced using the same hanger components as for noninsulated steel,

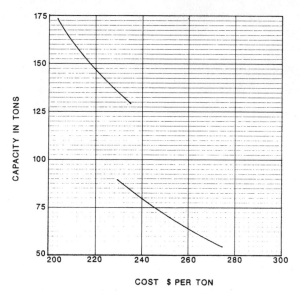

COST  $ PER TON

**Figure 7.15**  Air-cooled condenser assemblies.

except the clevis hanger was copper-plated and the piping was assumed to be hung from spring hangers and was priced that way.

Air-cooled condensers were assumed to be set on a base provided by other trades. Coil guards and starters were included. These prices are for a standard low ambient temperature of 40°. For lower-ambient-temperature assemblies you can add $800 for each air-cooled condenser.

### Cooling tower assemblies

The curves in Figure 7.16 show the installed costs in dollars per ton for the Marley series 4700, 47000, and 221 and the Baltimore Air Coil series VXT steel towers. We investigated the piping and labor costs for two typical locations: on the roof above the chiller and condenser and outside at the ground level at approximately the same elevation as the

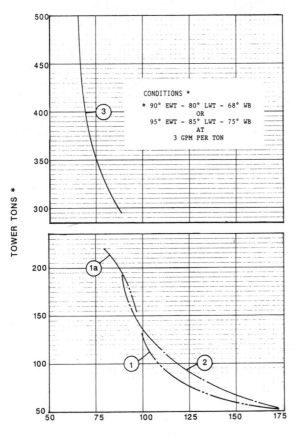

COST $ PER TON

**Figure 7.16**   Cooling tower assemblies. (1) Marley series 4700; (1a) Marley series 47000; (2) Baltimore Air Coil series VXT; (3) Marley series 221.

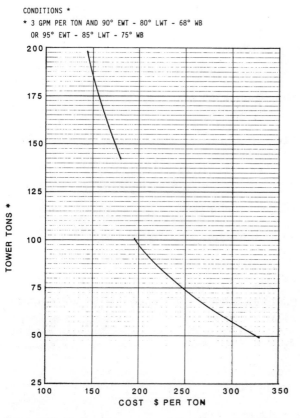

CONDITIONS *
* 3 GPM PER TON AND 90° EWT - 80° LWT - 68° WB
OR 95° EWT - 85° LWT - 75° WB

**Figure 7.17** Cooling tower assemblies: tower installed indoors.

chiller room floor. Costs were so nearly the same that we did not plot two curves.

Baltimore Air Coil tower costs were also plotted on Figure 7.17 for the situation where the tower is located indoors and air is ducted through the walls or through the wall and roof.

Since cooling tower capacities vary with the range (tower EWT − LWT) and the approach to the wet-bulb temperature, I have shown two different sets of conditions where the tower capacity is virtually the same. The first set of conditions can be used to approximate the tower's capacity in any of the dry country of the west and southwest. These conditions are 90° EWT − 80° LWT and 68° wet bulb (WB). The second set of conditions might be used in much of the rest of the United States and Canada except where extremely high humidity exists. These conditions are 95° EWT − 85° LWT and 75° WB.

All costs are complete installation costs including tower, motor, steel beams or concrete base supports, motor starter, material and piping costs a the condenser and cooling tower.

Piping at the condenser includes

2 gate valves

2 molded Teflon flexible connectors

2 flanges

2 temperature and pressure taps

2 spool pieces

2 elbows

Piping at the cooling tower includes

2 molded Teflon flexible connectors

2 flanges

2 roof or wall sleeves

2 butterfly valves installed indoors (furnished under the temperature control division)

4 companion flanges

2 tees

Water makeup piping includes

Dielectric union on makeup water

20 ft of steel makeup pipe

Screwed tee for plug or automatic drain

Roof or wall sleeve

Screwed elbow

Malleable iron union

2 nipples

Drain and overflow piping includes

4 nipples

2 unions

Roof or wall sleeves

3 screwed elbows

1 screwed tee for plug or automatic drain

1 gate valve on overflow

50 ft of drain piping

Additional costs for towers located indoors (Baltimore Air Coil series VXT) include

Rain louver on air intake

Automatic damper on air intake

5 ft of waterproof intake duct to tower

10 ft of waterproof discharge duct

Blow-open counter-balanced damper, and

Either a discharge rain louver or a duct flashed and counterflashed around curbed opening in the roof

Additional associated costs are the cost of chemical treatment (Table 7.6); freeze protection (Table 7.7); and pumping (see Figures 7.2 and 7.3). To these you must add piping costs between the condenser and cooling tower.

Tower water filtering may be a necessary or wise investment and should be investigated for each project and priced if needed.

### Cooling tower chemical treatment assemblies

We investigated and priced cooling tower water chemical treatment systems from Mogul, Chagrin Falls, Ohio. Table 7.6 shows the costs of three different treatment systems.

The first (and cheapest) system is the Mogul Solutrol II, which consists of automatic sampling, feeding of inhibitors, and automatic bleed. This system is usually used on smaller towers (under 200 tons) and where the makeup water is reasonably soft.

**TABLE 7.6  Cooling Tower Chemical Treatment Costs**

| Treatment system description | Tower makeup water size | Cost, $ Lower | Cost, $ Higher |
|---|---|---|---|
| Minimum treatment, automatic sampling, inhibitor feed, and bleed | ¾-in makeup | — | 1,500 |
| Automatic sampling, inhibitor and two biocides feed, and bleed | ¾- to 1-in makeup<br>1¼- to 1½-in makeup | 4,300<br>4,500 | 6,000<br>6,200 |
| Automatic sampling, inhibitor and two biocides, and acid feed with bleed | ¾- to 1-in makeup<br>1¼- to 1½-in makeup | 5,800<br>6,000 | 8,100<br>6,299 |

The second system is the Mogul Digital Auto-Chem D 62. This system should be used where makeup water is reasonably soft, not requiring acid feed. The system consists of automatic sampling, feeding of inhibitors plus two biocides, and automatic bleed in response to metered makeup water quantities.

The third system is the Mogul Digital Auto-Chem D 64, which is the same as the D 62 system plus the addition of acid feed. This system is for harder waters.

Two prices are shown for the last two systems. The higher cost is for the entire system as described in the Mogul literature. The lower cost is arrived at by deleting the locked enclosure for the Digital Auto-Chem chemical pumps and sample stream piping assembly (item 3.2 in Mogul specifications) and by deleting the liquid level switch assemblies, alarm lights and buzzers, and shutoff safety switches in the chemical drums (item 3.4 in Mogul specifications).

Costs include all mechanical contractor's work to install the system and the valves, etc., not furnished with the chemical treatment package, all as indicated on Mogul schematic drawings.

### Cooling tower freeze protection assemblies

Where outdoor temperatures and operating needs are encountered that require freeze protection of the tower piping and tower basin, we have priced one method of accomplishing this (see Table 7.7).

The method protects the makeup water pipe and drain and overflow pipe by wrapping them with heat tape and covering them with insulation and a weatherproof metal jacket. The heat tape is controlled from a contactor through a strap-on thermostat. The heat tape is interlocked to be off when the tower basin is drained.

Makeup water is provided to the tower through an automatic temperature control valve located inside the building where the normal tower float fill valve is deleted. This automatic valve is installed in a three-valve bypass with two unions and two tees.

TABLE 7.7   Cooling Tower Freeze Protection Costs

| Cooling tower description | Piping, $ | Basin heaters, $ | Total, $ |
| --- | --- | --- | --- |
| Up to 100 tons, ¾-in makeup, 2-in drain, 4-in piping | 2,070 | 965 | 3,030 |
| 101–300 tons, 1¼-in makeup, 2-in drain, 6-in piping | 2,220 | 985 | 3,205 |
| 301–500 tons, 1½-in makeup, 4-in drain, 8-in piping | 3,640 | 1,030 | 4,670 |

Connecting to the base of the makeup water line is an automatic drain valve with gate valve, union, two elbows, and piping connecting to the tower overflow drain line, which is to drain the outdoor piping. Connecting to the base of the tower drain line is an automatic drain valve with strainer, strainer blow-down valve, union, and piping to the overflow line. The gate valve on this drain line has previously been included in the normal tower costs.

Connecting to the base of each of the tower condenser water supply and return lines to drain the risers and outdoor piping is an assembly of a thread-o-let, two elbows, gate valve, strainer and strainer blow-down valve, union, automatic drain valve, nipples, and connection to the tower overflow drain.

All control valves, controls, and connecting wiring or pneumatic tubing are furnished under the temperature control division. The makeup water supply valve operates on demand from liquid level probes in the tower basin. When the tower basin is drained, the makeup water valve is closed. When tower water temperature approaches freezing conditions, all drain valves open and makeup water valve closes.

To protect the tower basin from freezing until the system is drained, an electric heater is installed. The system includes a weatherproof control box with low-voltage transformer and relay wired through a temperature control switch (operated from an immersion bulb) and a water level safety float switch to operate the electric heater to maintain the basin water temperature above freezing. All of these components and wiring are included. Power wiring to the control box is not included. For systems with heaters of 10 kW and above, it was assumed that 440 V power was available. For $7\frac{1}{2}$ kW and less, 208-V power was used to size the wiring.

### Chilled water and condenser water piping mains

Tables 7.1$a$ and $b$ (piping costs), and the discussion relating to same, apply as well to both chilled water and condenser water piping systems. Use those appropriate costs for cooling system mains from chiller room perimeter walls to large air-handling equipment and from condensers to cooling towers.

For perimeter cooling systems using small fan coils or unit ventilators we have drawn some theoretical floor plans and estimated the distribution mains for these systems; discussions follow. Also refer to the discussion under Fan Coil Assemblies and Unit Ventilator Assemblies.

**Exhibit 7.3**
**COOLING GENERATION AND DISTRIBUTION COST ESTIMATING SUMMARY SHEET**

|  | Total cost, $ |
|---|---|
| Chiller plant, type_____ |  |
| _____tons            @_____$/ton  = | _____ |
| Air-cooled condenser assembly_____tons × _____$/ton  = | _____ |
| Chilled water pump(s), series_____ |  |
| No. of pumps_____@_____ gpm × _____$/gpm  = | _____ |
| Pot feeder assembly: 2 gal      5 gal            = | _____ |
| Water feeder assembly            = | _____ |
| Antifreeze feeder assembly: Field-erect, factory-mounted            = | _____ |
| Cooling tower assembly, type_____ |  |
| _____tons            @_____$/ton  = | _____ |
| Chemical treatment assembly, type_____  = | _____ |
| Tower freeze protection, size_____tons  = | _____ |
| Tower water filtering assembly | _____ |
| Condenser water pump(s), series_____ |  |
| No. of pumps_____@_____ gpm × _____$/gpm  = | _____ |
| Condenser water mains: |  |
| _____ft            @_____$/ft  = | _____ |
| Miscellaneous_____ |  |
|   |   |
| Cooling portion estimate, this sheet | _____ |

Exhibit 7.3 is the Cooling Generation and Distribution Cost Estimating Summary Sheet, and we can enter the appropriate costs on it.

### Cost comparison curves for chilled water plants

To give an approximation of the comparative costs of complete chilled water plants we have plotted the curves on Figure 7.18. The water-cooled chiller costs include

Chiller and local piping

Chemical pot feeder assembly

Pumping assembly; one base-mounted pump

Cooling tower and local piping

Chemical treatment assembly

Tower freeze protection assembly

Tower water pumping assembly; one base-mounted pump

Tower piping, tower located 100-ft distant

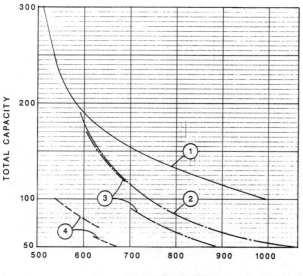

**Figure 7.18** Cost comparison curves for chilled water plants. (1) Hermetic centrifugal, water-cooled with tower; (2) hermetic reciprocating, water-cooled with tower; (3) hermetic reciprocating, condenserless with remote air-cooled condenser; (4) hermetic reciprocating air-cooled.

The air-cooled chiller costs include

Chiller and local piping

Antifreeze solution assembly

Pumping assembly; one base-mounted pump

The compressor chiller with remote air-cooled condenser costs include

Chiller and local piping

Chemical pot feeder assembly

Pumping assembly, one base-mounted pump

Air-cooled condenser assembly, including

Refrigerant piping from compressor-chiller to condenser

### Fan coil assemblies

Small fan coils come in sizes of 200, 300, 400, 600, 800, 1,000, and 1,200 cfm. The original family of fan coils was 200 through 600 cfm. In later years the 800-, 1,000-, and 1200-cfm family was developed. The larger units have a different price base and do not continue on the same curves applicable to the smaller sizes.

Three styles of fan coils were priced from the Trane Co.: (1) the

TABLE 7.8a   Fan Coil Assemblies Costs

| Style and size, cfm | 1. Two-pipe cooling, electric control | | 1a. Two-pipe changeover, electric control | | 2. Two-pipe cooling, pneumatic control | | 2a. Two-pipe changeover, pneumatic control | |
|---|---|---|---|---|---|---|---|---|
| | $ | $/cfm | $ | $/cfm | $ | $/cfm | $ | $/cfm |
| Vertical Cabinet Units | | | | | | | | |
| 200 | 832 | 4.16 | 852 | 4.28 | 780 | 3.90 | 795 | 3.98 |
| 300 | 857 | 2.86 | 877 | 2.92 | 800 | 2.67 | 815 | 2.72 |
| 400 | 883 | 2.21 | 903 | 2.26 | 826 | 2.07 | 841 | 2.10 |
| 600 | 920 | 1.53 | 940 | 1.57 | 863 | 1.44 | 878 | 1.46 |
| 800 | 1,263 | 1.58 | 1,283 | 1.60 | 1,186 | 1.48 | 1,201 | 1.50 |
| 1,000 | 1,338 | 1.34 | 1,358 | 1.36 | 1,261 | 1.26 | 1,276 | 1.28 |
| 1,200 | 1,441 | 1.20 | 1,461 | 1.22 | 1,364 | 1.14 | 1,379 | 1.15 |
| Horizontal Cabinet Units | | | | | | | | |
| 200 | 1,047 | 5.24 | 1,067 | 5.34 | 885 | 4.43 | 900 | 4.50 |
| 300 | 1,076 | 3.59 | 1,096 | 3.65 | 915 | 3.05 | 930 | 3.10 |
| 400 | 1,101 | 2.75 | 1,121 | 2.80 | 939 | 2.35 | 954 | 2.39 |
| 600 | 1,153 | 1.92 | 1,173 | 1.96 | 992 | 1.65 | 1,007 | 1.68 |
| 800 | 1,408 | 1.76 | 1,428 | 1.79 | 1,222 | 1.53 | 1,237 | 1.55 |
| 1,000 | 1,487 | 1.49 | 1,507 | 1.51 | 1,305 | 1.31 | 1,320 | 1.32 |
| 1,200 | 1,607 | 1.34 | 1,627 | 1.36 | 1,425 | 1.19 | 1,440 | 1.20 |
| Horizontal Concealed Units | | | | | | | | |
| 200 | 984 | 4.92 | 1,004 | 5.02 | 822 | 4.11 | 837 | 4.19 |
| 300 | 1,007 | 3.36 | 1,027 | 3.42 | 845 | 2.82 | 860 | 2.87 |
| 400 | 1,034 | 2.59 | 1,054 | 2.64 | 872 | 2.18 | 887 | 2.22 |
| 600 | 1,066 | 1.78 | 1,086 | 1.81 | 908 | 1.51 | 923 | 1.54 |
| 800 | 1,332 | 1.67 | 1,352 | 1.69 | 1,150 | 1.44 | 1,165 | 1.46 |
| 1,000 | 1,416 | 1.42 | 1,436 | 1.44 | 1,235 | 1.24 | 1,250 | 1.25 |
| 1,200 | 1,522 | 1.27 | 1,542 | 1.29 | 1,340 | 1.12 | 1,355 | 1.13 |

vertical cabinet model, (2) the horizontal cabinet model, and (3) the horizontal concealed model. Five different piping packages including temperature control valves, all factory-installed, were priced and are shown in Tables 7.8a and b; they are as follows:

1. One water coil with two-way control valve for two-pipe cooling only, with electric controls

2. One water coil with two-way control valve for two-pipe cooling only, with pneumatic controls

1a & 2a. One water coil with three-way control valve for automatic changeover system

3. One water coil with three-way control valve for two-pipe cooling and an auxiliary low-heat electric coil

**TABLE 7.8b   Fan Coil Assemblies Costs**

| Style and size, cfm | 3. Two-pipe, electric control with electric heating coil $ | $/cfm | 4. Four-pipe, electric control $ | $/cfm | 5. Four-pipe, pneumatic control $ | $/cfm | O.A. Package $ |
|---|---|---|---|---|---|---|---|
| | | | Vertical Cabinet Units | | | | |
| 200 | 1,209 | 6.05 | 1,359 | 6.80 | 1,250 | 6.25 | 185 |
| 300 | 1,240 | 4.13 | 1,381 | 4.60 | 1,303 | 4.34 | 200 |
| 400 | 1,269 | 3.17 | 1,409 | 3.52 | 1,331 | 3.33 | 220 |
| 600 | 1,314 | 2.19 | 1,454 | 2.42 | 1,376 | 2.29 | 220 |
| 800 | 1,792 | 2.24 | 1,854 | 2.32 | 1,695 | 2.12 | 255 |
| 1,000 | 1,897 | 1.90 | 1,941 | 1.94 | 1,782 | 1.78 | 270 |
| 1,200 | 2,043 | 1.70 | 2,054 | 1.71 | 1,895 | 1.60 | 300 |
| | | | Horizontal Cabinet Units | | | | |
| 200 | 1,321 | 6.61 | 1,560 | 7.80 | 1,355 | 6.78 | N/A |
| 300 | 1,354 | 4.51 | 1,600 | 5.33 | 1,396 | 4.65 | N/A |
| 400 | 1,382 | 3.46 | 1,622 | 4.06 | 1,417 | 3.54 | N/A |
| 600 | 1,441 | 2.40 | 1,682 | 2.80 | 1,477 | 2.46 | N/A |
| 800 | 1,829 | 2.29 | 1,994 | 2.49 | 1,765 | 2.21 | N/A |
| 1,000 | 1,937 | 1.94 | 2,085 | 2.09 | 1,856 | 1.86 | N/A |
| 1,200 | 2,101 | 1.75 | 2,215 | 1.85 | 1,986 | 1.66 | N/A |
| | | | Horizontal Concealed Units | | | | |
| 200 | 1,258 | 6.29 | 1,497 | 7.49 | 1,292 | 6.46 | N/A |
| 300 | 1,285 | 4.28 | 1,527 | 5.09 | 1,322 | 4.41 | N/A |
| 400 | 1,315 | 3.29 | 1,555 | 3.89 | 1,350 | 3.38 | N/A |
| 600 | 1,354 | 2.26 | 1,625 | 2.71 | 1,394 | 2.32 | N/A |
| 800 | 1,753 | 2.19 | 1,933 | 2.42 | 1,644 | 2.06 | N/A |
| 1,000 | 1,866 | 1.87 | 2,029 | 2.03 | 1,740 | 1.74 | N/A |
| 1,200 | 2,016 | 1.68 | 2,145 | 1.79 | 1,857 | 1.55 | N/A |

4. Two water coils with three-way control valves for four-pipe systems, with electric controls

5. Two water coils with three-way control valves for four-pipe systems, with pneumatic controls

Table 7.8b also shows the cost of a 0 to 25% outside air package with automatic control damper and anodized aluminum wall box for use with vertical units; add this cost to the basic unit cost where applicable.

Prices include baked-enamel finish for cabinet models, an enclosure for the fan on the horizontal concealed model, 1-in glass fiber throw-away filters, standard motors, and 15-ft-long piping runouts without valves (valves included in units). Transformers (based on a 208/277-V power supply) to 120 V were included on models with electric coils; all others were priced on 120-V power. Also included were factory-mounted valve packages including control valves, shutoff valves, balancing valve, and air vent. Where Trane did not mount the control valve, we added the cost of field mounting the valve.

For units with electric control, the following control packages were used (control wiring costs have been included):

**Vertical models**

1. One water coil for cooling only: motor switch, thermostat, and manual cooling-ventilating switch
2. One water coil for two-pipe changeover: motor switch, thermostat, and automatic summer-winter switch
3. One water coil for cooling with auxiliary electric heating coil: motor switch, thermostat, and manual summer-winter switch
4. Two water coils for four-pipe system: motor switch, two-stage thermostat, and automatic summer-winter switch

**Horizontal models.** All controls were the same except that they were wall mounted in the field. Control wiring has been included in these prices.

For units with pneumatic control you must add the following costs in the control section:

Price of the valves

Thermostat and tubing, material and labor costs

Summer-winter control, material and labor costs

Price of connecting outside air damper motor

For all units you must add the cost of drain pan piping.

Because you almost always end up with a room air requirement somewhat less than the fan coil you will pick for the room, this system is inefficient in matching the equipment to the load. To estimate the number of fan coils required, take your calculated air volume and divide by .85 to get a theoretical air load to match with your fan coils. From this cubic-foot-per-minute rate, make an estimate of the number of fan coils of each type and size and estimate their costs from Tables 7.8a and b.

**Unit ventilator assemblies**

Unit ventilators were priced using Trane Co. equipment. All prices include a baked-enamel cabinet, normal-capacity coils, 1-in glass fiber

**TABLE 7.9 Unit Ventilator Assemblies**

| Style and size, cfm | 1. Heating with valve control | | 2. Heating and cooling, face and bypass, two-pipe changeover | | 3. Heating and cooling, valve control, four-pipe | |
|---|---|---|---|---|---|---|
| | $ | $/cfm | $ | $/cfm | $ | $/cfm |
| Vertical Cabinet | | | | | | |
| 750 | 1,666 | 2.22 | 1,851 | 2.47 | 2,321 | 3.09 |
| 1,000 | 1,790 | 1.79 | 1,980 | 1.98 | 2,466 | 2.47 |
| 1,250 | 1,928 | 1.54 | 2,101 | 1.68 | 2,756 | 2.20 |
| 1,500 | 2,063 | 1.38 | 2,369 | 1.58 | 2,930 | 1.95 |
| Horizontal Cabinet | | | | | | |
| 750 | 1,803 | 2.40 | 1,980 | 2.64 | 2,488 | 3.32 |
| 1,000 | 1,899 | 1.90 | 2,067 | 2.07 | 2,597 | 2.60 |
| 1,250 | 2,064 | 1.65 | 2,331 | 1.86 | 2,929 | 2.34 |
| 1,500 | 2,208 | 1.47 | 2,498 | 1.67 | 3,123 | 2.08 |

throwaway filters, a wall box with anodized grille (arrangement 2), stop valves, and factory-piped and -mounted balancing valve and control valve. Where the Trane Co. does not mount the control valve, the valve cost should be entered into the temperature control cost; we added the cost of field mounting the valve. Also included were 15-ft runouts without valves.

Both vertical and horizontal units were priced, each with three-coil and control combinations; they were

1. One water coil with valve control for heating only

2. One water coil with face and bypass control for two-pipe changeover system

3. Two water coils with valve control for four-pipe systems

All costs in Table 7.9 are for pneumatic controls. To these prices you must add the following costs in the control section:

Price of the valves

Thermostat and tubing, material and labor costs

Summer-winter control, material and labor costs

Price of connecting outside air damper motor

For cooling units you must add the cost of drain pan piping.

## Perimeter piping mains for fan coils and unit ventilators

To investigate cooling loads for typical buildings where fan coils might be employed, we reviewed the past projects of McFall-Konkel & Kimball. It appeared that fan coils could be employed for perimeter cooling and heating in the following types of buildings*:

| | |
|---|---|
| Office buildings | Apartments |
| Hospital patient rooms | Condominiums |
| Nursing homes | Dormitory student rooms |
| Motel guest rooms | Fraternity-sorority student rooms |
| Hotel guest rooms | |

To estimate the loads for buildings with fan coils and unit ventilators, we used the data shown in Table 7.10 as input. Loads were calculated at Denver, Colorado, using $71°dt$, outdoor temperature to room temperature for winter heating. For cooling we used the 97.5% column for outside dry-bulb temperature from the ASHRAE fundamentals. Rooms peak in different months, depending on orientation. We used an inside room temperature of 75° with 55° air off of the coil.

Building floor-to-floor height was set at 12 ft with 50% insulated panel and 50% double-pane glass.

We arrived at loads which would lead us to some generalizations. First we expanded the office width to 12.5 ft wide to match other room types. Doing this we could use the same piping arrangement for all room types using fan coils. Sizes of fan coils and unit ventilators for each room vary with room outside wall orientation. Cubic feet per minute calculated for each room was rounded up to the next size fan coil or unit ventilator.

Mains serving one floor from the ceiling space above and mains serving two floors from the ceiling space between the floors were sized and priced. These curves are for end-fed systems as shown in Figure 7.19. Costs can be greatly reduced if the piping can be center-fed for either two or four building wings.

To read the curves in Figures 7.20 to 7.23 find the number of fan coils or unit ventilators for the building wing, intersect the appropriate curve, and read the cost in dollars per unit at the left. For center-fed two-building-wing systems, this will be the cost in dollars per unit for two times the number of units for end-fed. For center-fed four-building-

---

* Buildings typically of a rectangular shape with rooms on each side of a corridor. Fan coils are to handle the perimeter only; the core of the building is to be handled by some other system.

**TABLE 7.10  Cooling Load Data**

| Building type | Room size | | U factors | | | | Glass shading coefficient | W/ft² | People per room | Outside air, cfm/room |
|---|---|---|---|---|---|---|---|---|---|---|
| | Wall length, ft | Depth, ft | Wall | Glass* Us | Glass* Uw | Roof | | | | |
| Office | 10 | 12.5 | .09 | .57 | .55 | .05 | .50 | 3 | 1 | 50 |
| Hospital | 12.5 | 16 | .09 | .57 | .55 | .05 | .50 | 2 | 2 | 50 |
| Nursing home | 12.5 | 16 | .09 | .57 | .55 | .05 | .50 | 2 | 2 | 50 |
| Motel, hotel | 12.5 | 20 | .09 | .57 | .55 | .05 | .50 | 1.5 | 2 | 50 |
| Apartments, condominiums | 12.5 | 16 | .09 | .57 | .55 | .05 | .50 | 1.5 | 1 | 50 |
| Dormitory rooms, fraternity-sorority houses | 12.5 | 16 | .09 | .57 | .55 | .05 | .50 | 1.5 | 2 | 50 |
| Classroom | 35 | 21 | .09 | .57 | .55 | .05 | .50 | 3.0 | 25 | 187† |

* $U_s$ = U value for summer; $U_w$ = U value for winter.
† 7.5 cfm per person.

113

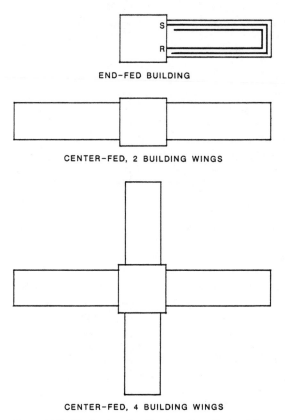

Figure 7.19  Perimeter heating and cooling piping layouts.

wing systems, this will be the cost in dollars per unit for four times the number of units for end-fed systems. The economy of center-fed systems can easily be seen from these numbers.

### Sizing fan coil piping equipment

For east-west orientation,
   All building types except offices: 400 cfm, each room both east and west
   Offices: 400 cfm, west rooms; 300 cfm, east rooms

For north-south orientation,
   All building types: 400 cfm, south rooms; 300 cfm, north rooms

Building wings with lengths of 75, 100, 150, and 200 ft were used to lay out a piping system for each.

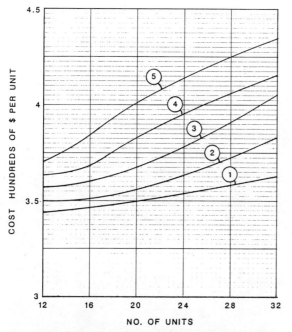

**Figure 7.20**  Fan coil perimeter heating and cooling piping mains cost, mains serving one floor, end-fed. (1) 200-cfm fan coils each side of corridor, also heating piping main costs for four-pipe systems for 70°$dt$ climates at 20° water temperature drop; (2) 200-cfm fan coils one side of corridor, 300-cfm fan coils the other side of corridor; (3) 300-cfm fan coils each side of corridor, also heating piping main costs for four-pipe systems for 100°$dt$ climates at 20° water temperature drop; (4) 300-cfm fan coils one side of corridor, 400-cfm fan coils the other side of corridor; (5) 400-cfm fan coils each side of corridor.

### Sizing unit ventilator piping equipment

For east-west orientation: 1,500 cfm, each room both east and west

For north-south orientation: 1,500 cfm, south rooms; 1,250 cfm, north rooms

Building wings of 105, 140, and 210 ft were used to lay out a piping system for each.

### AC and HV units

Numerous options are available on air conditioning (AC) and HV units, which will vary the costs shown on Figure 7.24. Our standards, which Figure 7.24 was based on, were as follows:

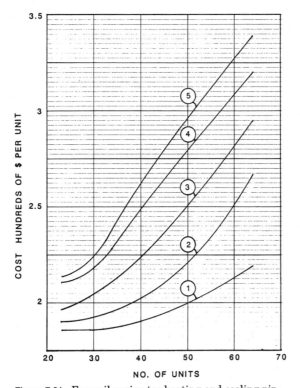

**Figure 7.21**  Fan coil perimeter heating and cooling piping mains cost, mains serving two floors, end-fed. (1)
200-cfm fan coils each side of corridor, also heating
piping main costs for four-pipe systems for 70°*dt* climates at 20° water temperature drop; (2) 200-cfm fan
coils one side of corridor, 300-cfm fan coils the other side
of corridor; (3) 300-cfm fan coils each side of corridor,
also heating piping main costs for four-pipe systems for
100°*dt* climates at 20° water temperature drop; (4)
300-cfm fan coils one side of corridor, 400-cfm fan coils
the other side of corridor; (5) 400-cfm fan coils each side
of corridor.

## HV unit: No coils

670 to 710 ft/min face velocity and 2 in total static pressure (TSP)

Horizontal draw-through unit

Standard casing with access door to fan section o9r fan and coil
section on vertical units

Combination filter-mixing box with low-leak dampers both openings

2-in glass fiber throwaway filters

1-in, 1½-lb insulation, all components

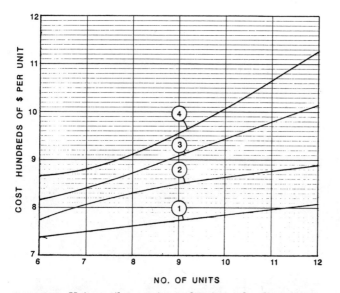

**Figure 7.22**    Unit ventilator perimeter heating and cooling piping mains cost, mains serving one floor, end-fed. (1) Heating mains cost, climates with $70°dt$; (2) heating mains cost, climates with $115°dt$, also cooling mains cost for 1,000-cfm unit ventilators; (3) cooling mains cost, 1,250-cfm unit ventilators; (4) cooling mains cost, 1,500-cfm unit ventilators.

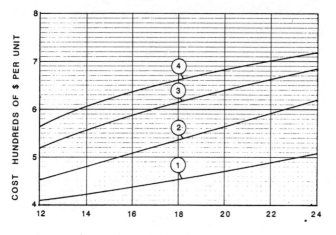

**Figure 7.23**    Unit ventilator perimeter heating and cooling piping mains cost, mains serving two floors, end-fed. (1) Heating mains cost, climates with $70°dt$; (2) heating mains cost, climates with $115°dt$, also cooling mains cost for 1,000-cfm unit ventilators; (3) cooling mains cost, 1250-cfm unit ventilators; (4) cooling mains cost, 1,500-cfm unit ventilators.

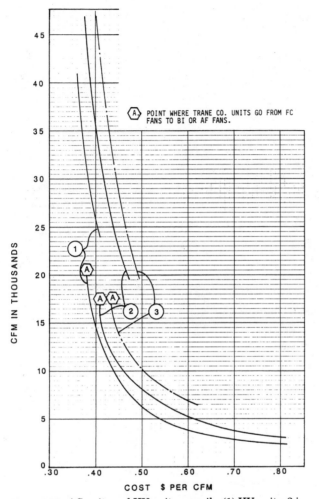

**Figure 7.24** AC units and HV units, no coils. (1) HV units, 2 in total static pressure (TSP); (2) AC units, 2½ in TSP; (3) AC units, 4½ in TSP.

Motor and starter for 208/230 V—increment start, 30 hp and above

Adjustable V-belt drive and standard belt guard

Isolation base, spring isolators, and 4-in concrete pad

Air foil wheels on units size 35 and above

### AC units: No coils

543 to 558 ft/min face velocity

Low-pressure units at $2\frac{1}{2}$ in TSP

TABLE 7.11    Optional Costs for AC and HV Units

| Unit size | HV unit, cfm @ selection | AC unit, cfm @ selection | Vertical unit, AC only, ¢/cfm | 4-in permanent filters, ¢/cfm | | Inlet vanes, medium pressure, AC only, ¢/cfm | External face and bypass with duct, ¢/cfm | |
|---|---|---|---|---|---|---|---|---|
| | | | | HV | AC | | HV | AC |
| 6 | 3,800 | 3,080 | +2 | — | — | — | +8 | +10 |
| 8 | 4,900 | 4,125 | +2 | — | — | — | +8 | +9 |
| 10 | 6,000 | 5,170 | +1 | — | — | — | +7 | +8 |
| 12 | 7,800 | 6,500 | +1 | +3 | +4 | +12 | +6 | +7 |
| 14 | 9,000 | 7,500 | 0 | +3 | +4 | +10 | +6 | +7 |
| 17 | 11,000 | 9,200 | 0 | +3 | +3 | +9 | +5 | +6 |
| 21 | 14,000 | 11,500 | +2 | +2 | +3 | +8 | +5 | +6 |
| 25 | 16,400 | 13,200 | +3 | +2 | +2 | +7 | +5 | +6 |
| 31 | 20,000 | 16,800 | +2 | +1 | +1 | +6 | +5 | +6 |
| 35 | 24,000 | 19,500 | — | 0 | 0 | +6 | +5 | +6 |
| 41 | 27,500 | 22,000 | — | 0 | 0 | +6 | +5 | +6 |
| 50 | 33,700 | 27,000 | — | 0 | 0 | +5 | +5 | +6 |
| 63 | 41,000 | 34,000 | — | 0 | 0 | +4 | +4 | +5 |
| 73 | — | 40,000 | — | — | — | +3 | — | +5 |
| 86 | — | 47,000 | — | — | — | +3 | — | +5 |

Medium-pressure units at $4\frac{1}{2}$ in TSP

Horizontal draw-through unit, arrangement 1, 2, 3, or 4 for sizes up through size 21; above this size the prices based on arrangement 1 or 2

Arrangement 4 costs from $500 more at size 25 to $1,500 more at size 63; arrangement 3 costs from $500 more at size 25 to $1,000 more at size 63. All other standards are the same as for HV units.

Table 7.11 shows the cubic-feet-per-minute selection points for HV and AC units and also shows the costs of several options. Vertical AC units cost more than horizontal AC units and also take more horsepower per cubic foot per minute to move the air. The cost of 4-in permanent filters entailed deleting the combination filter and mixing box and 2-in filters and adding a separate mixing box and filter box with 4-in filters. The added cost is small and for the larger sizes reduces to practically zero. At size 35 and larger, separate mixing and filter boxes cost less than a combination filter and mixing box.

The added cost of inlet vanes is applicable only to medium-pressure AC unit cost. We investigated other options. Hanging units with spring isolators rather than mounting on the floor costs more for the labor but deletes the isolation base and 4-in concrete pad. The cost is the same either way.

Two-inch permanent filters cost $.015/cfm more than 2-in throwaway filters for the smallest unit to $.01/cfm more for the largest unit.

Low-leak dampers add $.02/cfm more for the smallest unit to $.01/cfm more for the largest unit if they are used in both mixing box openings. They were included in our standard.

Air foil fans, which we chose as standard, cost $.015/cfm more than backward-inclined fans at unit size 35. This cost diminishes to $.01/cfm at the larger sizes.

We priced blow-through units, and they are generally more expensive than draw-through units, although no predictable pattern of pricing in cost per cubic foot per minute could be found. Some unit sizes cost as much as $.05/cfm more than draw-through units; others were essentially the same price. However, it is interesting to note that while draw-through units add the heat of compression from the fan to the *already* cooled air, thus requiring subcooling of the air stream and the resulting cost of same, blow-through units require more horsepower per cubic foot per minute. (They also add the heat of compression to the air stream, but before it is cooled.) Our figures show that at $2\frac{1}{2}$ in TSP the blow-through units require about $\frac{1}{2}$ hp per 10,000 cfm more for the smallest unit, ranging up to $1\frac{3}{4}$ hp per 10,000 cfm for the largest unit. At $4\frac{1}{2}$ in TSP these numbers increase to 1 hp per 10,000 cfm and $2\frac{1}{2}$ hp per 10,000 cfm. The economics for any given size unit could be calculated comparing the two when design is started.

To the cost of all these units you must add the cost of coils and coil piping.

### Multizone and double-duct units

Most of the same options that are available for single-zone AC and HV units are available on multizone and double-duct units. Figure 7.25 shows the cost of HV multizones, two- and three-deck multizones, and double-duct units. As with single-zone units, the multizones get more expensive in cost per cubic foot per minute, where the design goes from forward-curved fan wheels to backward-inclined or air foil (AF) fan wheels. An interesting point, however, is that the next two sizes do not get cheaper in cost per cubic foot per minute as is the usual case in equipment. This can be seen from the reverse curves in Figure 7.25 from 19,500 to 34,000 cfm.

Our standards for pricing the multizone units were the same as for single-zone units. The cooling multizone cubic-feet-per minute rate was selected at $2\frac{1}{2}$ in TSP. The costs for options such as permanent filters, low-leak dampers, air foil fans, and 4-in filters are the same as for single-zone AC units.

### Coil assemblies

Coils vary in configuration by size, square to rectangular shape, and by rows, fin type, and fin spacing. Selection varies as well by entering

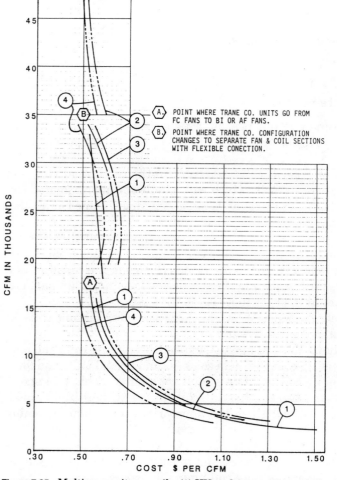

**Figure 7.25**  Multizone units, no coils. (1) HV multizone; (2) two-deck multizone; (3) three-deck multizone; (4) double-duct unit.

water temperature, water flow, and sensible to total heat ratio. There are so many combinations and variations in coil design and selection that it was an impossible task to price them all. Coils shown here were selected as follows.

### General

Fan cubic feet per minute rate was selected based on building load. Coil capacity was selected based on total load, including building load plus 25% of that load representing outdoor air load, fan heat of compression, etc.

Trane Co. Type W water coils (Prima fin type was chosen for lower fan static pressure drop).

### Cooling for four- and six-row coils,

42° EWT, 2.4° gpm/ton

80° EAT, 55° ± LAT

550 ft/min face velocity

**Heating.** One- and two-row heating coils were not sized for capacity. Whether to use one- or two-row coils is left to the judgment of the estimator.

A great deal of experience is necessary in the conceptual stages of design in order to select coils which will handle the approximate load. It is expected that the designer has the judgment gained from this experience to make reasonable selections.

Coil prices here were divided into coils within units and field-erected coils or coil banks. For field-erected coil banks we assumed the Trane Co. coil module was a reasonable method of housing the coils through their size range, rather than field erecting the housing, drain pan, etc. From our pricing it appears that this is the case.

Table 7.12 shows the cost of one- and two-row heating coils in single-zone and multizone HV units.

Table 7.13 shows the cost of four- and six-row cooling coils and one- and two-row heating coils in single-zone AC and HVAC units and four- and six-row cooling coils in HVAC multizone units. Table 7.14 shows the cost of one- and two-row heating coils in HVAC multizone units.

To these costs you must add the appropriate coil piping assembly cost.

Table 7.15 shows the cost of the Trane Co. coil modules when used as a separate entity. To these costs, you must add the cost of coils from Table 7.13 and the appropriate coil piping assembly cost.

Figure 7.26 shows the components used in pricing two- and three-high coil banks in sizes larger than the coil module will accommodate.

Figure 7.27 is a set of curves showing costs of field-erected six-row cooling and two-row heating coils for conceptual estimating, if an accurate grasp of the number of rows is not available.

Table 7.16 is a chart showing the cost of four- and six-row cooling coils and one- and two-row heating coils for field-erected banks of coils.

To these costs you must add the appropriate coil piping assembly cost.

Coils are priced with a base coil cost plus a cost factor times the number of fins per foot. Our selections were made so that as the actual number of fins per foot varies from those chosen, the total coil price should not vary more than 8%, with 5% variation being an average.

TABLE 7.12 Coil Costs for HV Units (Coils in Units Only)

| Unit size | Full coil ft² | Full coil Size | cfm @ selection | Heating, $ IR-W-140-P | Heating, $ 2R-W-110-P | Multizone ft² | Multizone Size | Heating, $ IR-W-150-P | Heating, $ 2R-W-120-P |
|---|---|---|---|---|---|---|---|---|---|
| 6 | 5.3 | 18 × 42 | 3,800 | 293 | 389 | 3.5 | 12 × 42 | 241 | 297 |
| 8 | 6.9 | 30 × 33 | 4,900 | 391 | 477 | 5.5 | 24 × 33 | 347 | 412 |
| 10 | 8.8 | 30 × 42 | 6,000 | 421 | 529 | 7.0 | 24 × 42 | 371 | 458 |
| 12 | 11.0 | 33 × 48 | 7,800 | 461 | 615 | 8.0 | 24 × 48 | 389 | 500 |
| 14 | 13.1 | 33 × 67 | 9,000 | 509 | 675 | 9.5 | 24 × 57 | 427 | 550 |
| 17 | 16.3 | 30 × 78 | 11,000 | 559 | 757 | 13.1 | 24 × 78 | 490 | 660 |
| 21 | 20.0 | 30 × 96 | 14,000 | 622 | 866 | 16.0 | 24 × 96 | 542 | 752 |
| 25 | 23.4 | 33 × 102 | 16,400 | 675 | 975 | 17.0 | 24 × 102 | 562 | 784 |
| 31 | 29.8 | 42 × 102 | 20,000 | 992 | 1,401 | 23.4 | 33 × 102 | 688 | 1,000 |
| 35 | 34.0 | 2-24 × 102 | 24,000 | 1,103 | 1,496 | 25.5 | 2-18 × 102 | 902 | 1,281 |
| 41 | 39.4 | 1-24 × 105 1-30 × 105 | 27,500 | 1,217 | 1,670 | 30.6 | 1-18 × 105 1-24 × 105 | 1,035 | 1,434 |
| 50 | 48.2 | 2-33 × 105 | 33,700 | 1,367 | 1,940 | 37.2 | 1-18 × 105 1-33 × 105 | 1,172 | 1,648 |
| 63 | 61.2 | 2-30 × 105 1-24 × 105 | 41,000 | 1,871 | 2,570 | 48.2 | 2-33 × 105 | 1,393 | 1,993 |

NOTE: Abbreviations: 1R = 1 row, W = coil type, 140 = fin spacing, P = Prima fins.

**TABLE 7.13  Coil Costs for AC and HVAC Units (Coils in Units Only)**

| Unit size | Full coil ft² | Full coil Size | cfm @ selection | Cooling 6R-W-110-P $ | Cooling 6R-W-110-P $/ft² | Cooling 4R-W-120-P $ | Cooling 4R-W-120-P $/ft² | Heating 1R-W-140-P $ | Heating 1R-W-140-P $/ft² | Heating 2R-W-110-P $ | Heating 2R-W-110-P $/ft² |
|---|---|---|---|---|---|---|---|---|---|---|---|
| 6 | 5.6 | 18 × 45 | 3,080 | 824* | 147 | — | — | 287 | 51 | 384 | 69 |
| 8 | 7.5 | 30 × 36 | 4,125 | 1,042* | 139 | — | — | 380 | 51 | 468 | 62 |
| 10 | 9.4 | 30 × 45 | 5,170 | 1,230* | 131 | — | — | 413 | 44 | 524 | 56 |
| 12 | 11.7 | 33 × 51 | 6,500 | 1,427* | 122 | — | — | 455 | 39 | 602 | 51 |
| 14 | 13.8 | 33 × 60 | 7,500 | 1,629* | 123 | — | — | 495 | 36 | 658 | 48 |
| 17 | 16.9 | 30 × 81 | 9,200 | 1,687 | 100 | 1,369* | 81 | 542 | 32 | 739 | 44 |
| 21 | 20.6 | 30 × 99 | 11,500 | 1,975 | 96 | 1,591* | 77 | 601 | 29 | 841 | 41 |
| 25 | 24.1 | 33 × 105 | 13,200 | 2,233 | 93 | 1,791* | 74 | 650 | 27 | 938 | 39 |
| 31 | 30.6 | 1-24 × 105<br>1-18 × 105 | 16,800 | 3,053 | 100 | 2,473* | 81 | 962 | 31 | 1,352 | 44 |
| 35 | 35.0 | 2-24 × 105 | 19,500 | 3,338 | 95 | 2,747* | 78 | 1,066 | 30 | 1,446 | 41 |
| 41 | 40.5 | 1-24 × 108<br>1-30 × 108 | 22,000 | 3,744 | 92 | 3,021* | 75 | 1,184 | 29 | 1,607 | 40 |
| 50 | 49.5 | 2-33 × 108 | 27,000 | 4,480 | 90 | 3,600* | 73 | 1,333 | 22 | 1,874 | 38 |
| 63 | 63.0 | 1-24 × 108<br>2-30 × 108 | 34,000 | 5,827 | 92 | 4,679* | 74 | 1,792 | 28 | 2,476 | 39 |
| 73 | 72.5 | 3-30 × 116 | 40,000 | 6,929 | 96 | 5,505* | 76 | 1,986 | 27 | 2,848 | 39 |
| 86 | 86.3 | 3-30 × 138 | 47,000 | 7,927 | 92 | 6,283* | 73 | 2,212 | 26 | 3,250 | 38 |

*With turbulators.

NOTE: Abbreviations: 1R = 1 row, W = coil type, 140 = fin spacing, P = Prima fins.

TABLE 7.14    Heating Coils Costs for HVAC Multizone Units (Coils in Units Only)

| Unit size | Blow-through hot deck ft² | Size | Cost of heating, $ 1R-W-150-P | 2R-W-120-P |
|---|---|---|---|---|
| 6 | 3.0 | 12 × 36 | 218 | 262 |
| 8 | 3.8 | 18 × 30 | 255 | 327 |
| 10 | 4.9 | 18 × 39 | 276 | 354 |
| 12 | 6.0 | 18 × 48 | 294 | 407 |
| 14 | 7.1 | 18 × 57 | 326 | 445 |
| 17 | 8.6 | 18 × 69 | 351 | 484 |
| 21 | 10.5 | 18 × 84 | 384 | 546 |
| 25 | 12.4 | 18 × 99 | 421 | 593 |
| 31 | 15.5 | 24 × 93 | 505 | 694 |
| 35 | 17.0 | 24 × 102 | 531 | 725 |
| 41 | 20.6 | 30 × 99 | 612 | 847 |
| 50 | 24.1 | 33 × 105 | 663 | 943 |
| 63 | 30.6 | 18 × 105 | 977 | 1,365 |
|  |  | 24 × 105 |  |  |
| 73 | 37.7 | 2-24 × 113 | 1,012 | 1,594 |
| 86 | 45.0 | 2-24 × 135 | 1,280 | 1,836 |

NOTE: Abbreviations: 1R = 1 row, W = coil type, 150 = fin spacing, P = prima fins.

## Coil piping assemblies

Figures 7.28 through 7.32 show the piping assemblies for the coil arrangements used for estimating the piping costs. Fifteen feet of piping to the mains was always used. Connections to the mains are included in the cost of the mains and are not included here. All coil

TABLE 7.15    Coil Modules Cost* When Used as a Separate Entity

| Unit size | Face area, ft² | cfm @ selection point | Total cost, $ | Cost, $/ft² |
|---|---|---|---|---|
| 06 | 5.6 | 3,080 | 733 | 131 |
| 08 | 7.5 | 4,125 | 830 | 111 |
| 10 | 9.4 | 5,170 | 867 | 92 |
| 12 | 11.7 | 6,500 | 913 | 78 |
| 14 | 13.8 | 7,500 | 943 | 68 |
| 17 | 16.9 | 9,200 | 1,211 | 72 |
| 21 | 20.6 | 11,500 | 1,284 | 62 |
| 25 | 24.1 | 13,200 | 1,386 | 58 |
| 31 | 30.6 | 16,800 | 1,679 | 55 |
| 35 | 35 | 19,500 | 2,371 | 68 |
| 41 | 40.5 | 22,000 | 2,458 | 61 |
| 50 | 49.5 | 27,000 | 2,688 | 54 |
| 63 | 63 | 34,000 | 2,833 | 45 |

* Add the cost of appropriate coils to this cost. This cost is for the cabinet, drain pan, and base only. Use the costs of coils from Figure 7.13. These are costs of coils installed in the cabinet at the factory. They are different from the cost of field-erected coil banks.

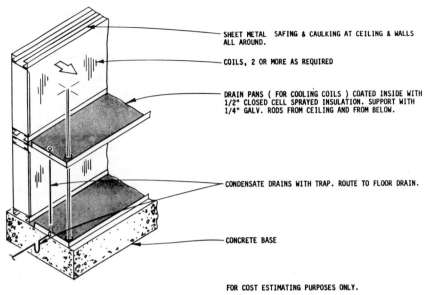

SHEET METAL SAFING & CAULKING AT CEILING & WALLS ALL AROUND.

COILS, 2 OR MORE AS REQUIRED

DRAIN PANS ( FOR COOLING COILS ) COATED INSIDE WITH 1/2" CLOSED CELL SPRAYED INSULATION. SUPPORT WITH 1/4" GALV. RODS FROM CEILING AND FROM BELOW.

CONDENSATE DRAINS WITH TRAP. ROUTE TO FLOOR DRAIN.

CONCRETE BASE

FOR COST ESTIMATING PURPOSES ONLY.

NO SCALE

**Figure 7.26**  Field-erected coil banks.

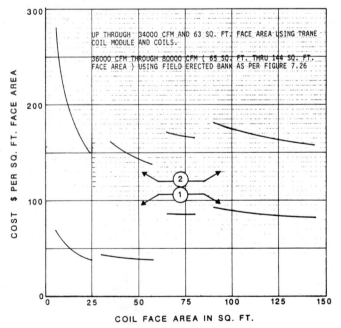

**Figure 7.27**  Field-erected coil banks cost. (1) **Two-row heating**; (2) **six-row cooling**.

TABLE 7.16    Field-Erected Coil Banks Cost

| Flow, cfm | Face area, ft² | No. of coils and coil sizes | Cost of 6-row cooling, $ | Cost of 4-row cooling, $ | Cost of 2-row heating, $ | Cost of 1-row heating, $ |
|---|---|---|---|---|---|---|
| 3,080 | 5.6 | 06 module | 1,557 | — | 384 | 287 |
| 4,125 | 7.5 | 08 module | 1,872 | — | 468 | 380 |
| 5,170 | 9.4 | 10 module | 2,097 | — | 524 | 413 |
| 6,500 | 11.7 | 12 module | 2,340 | — | 602 | 455 |
| 7,500 | 13.8 | 14 module | 2,572 | — | 658 | 495 |
| 9,200 | 16.9 | 17 module | 3,186 | 2,580 | 739 | 542 |
| 11,500 | 20.6 | 21 module | 3,259 | 2,875 | 841 | 601 |
| 13,200 | 24.1 | 25 module | 3,619 | 3,177 | 938 | 650 |
| 16,800 | 30.6 | 31 module | 4,732 | 4,152 | 1,352 | 962 |
| 19,500 | 35.0 | 35 module | 5,709 | 5,118 | 1,446 | 1,066 |
| 22,000 | 40.5 | 41 module | 6,202 | 5,479 | 1,607 | 1,184 |
| 27,000 | 49.5 | 50 module | 7,168 | 6,288 | 1,874 | 1,333 |
| 34,000 | 63.0 | 63 module | 8,666 | 7,512 | 2,476 | 1,792 |
| 36,000 | 65 | 1-48 × 120 1-30 × 120 | 11,158 | 9,251 | 5,589 | 1,681 |
| 42,000 | 80 | 2-48 × 120 | 13,286 | 10,968 | 6,840 | 1,799 |
| 48,000 | 90 | 3-36 × 120 | 15,999 | 13,371 | 8,154 | N/A |
| 54,000 | 105 | 3-42 × 120 | 18,600 | 15,339 | 9,645 | N/A |
| 60,000 | 110 | 2-42 × 120 1-48 × 120 | 19,043 | 15,710 | 9,850 | N/A |
| 66,000 | 120 | 3-48 × 120 | 19,929 | 16,452 | 10,260 | N/A |
| 72,000 | 132 | 2-42 × 144 1-48 × 144 | 21,450 | 17,871 | 11,062 | N/A |
| 80,000 | 144 | 3-48 × 144 | 22,254 | 18,729 | 11,532 | N/A |

connections are assumed to be screwed connections, $2\frac{1}{2}$ in being the largest size.

Shutoff valves at the pump are assumed to suffice also for the coils. For pumped coils, see Figures 7.2 or 7.3 for pump costs to add to these costs. Also add the costs of primary-secondary piping bridles (see Table 7.2) if applicable. Table 7.17 shows the costs of piping the different coils investigated. Add the cost of the coils to these costs.

### Flexible connectors

Where coils are in air-handling units that are supported on springs or other vibration isolators, add the cost of flexible connectors in the piping at the coil (see Table 7.1b).

### Summary

Exhibit 7.4 is the  summary sheet for cooling, heating, and HVAC equipment and piping.

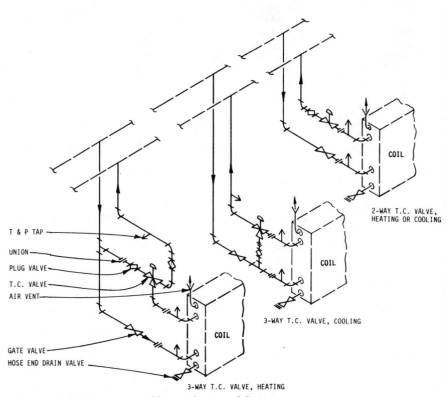

**Figure 7.28**  Coil piping assemblies with screwed fittings.

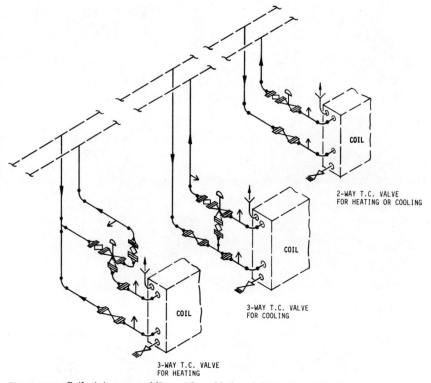

**2-WAY T.C. VALVE FOR HEATING OR COOLING**

**3-WAY T.C. VALVE FOR COOLING**

**3-WAY T.C. VALVE FOR HEATING**

**Figure 7.29** Coil piping assemblies with welded and flanged construction.

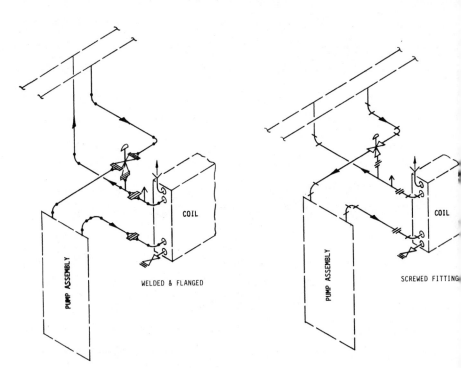

**Figure 7.30**  Pumped coil piping assemblies, cooling.

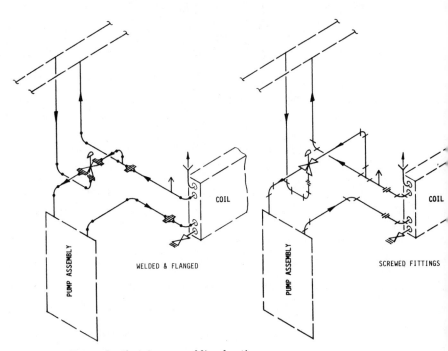

**Figure 7.31**  Pumped coil piping assemblies, heating.

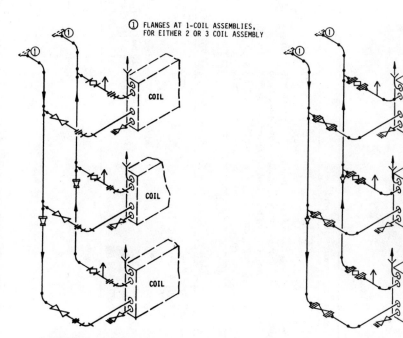

① FLANGES AT 1-COIL ASSEMBLIES,
FOR EITHER 2 OR 3 COIL ASSEMBLY

SCREWED BRANCHES & WELDED HEADERS

WELDED & FLANGED CONSTRUCTION

**Figure 7.32** Multiple coil piping assemblies.

TABLE 7.17  Coil Piping Assemblies Cost

| | 1. Two-way temperature control valve for either cooling or heating | | 2. Three-way bypass temperature control valve for cooling | | 3. Three-way bypass temperature control valve for heating | | 4. Three-way mixing temperature control valve for cooling (pumped coil) | | 5. Three-way mixing temperature control valve for heating (pumped coil) | | 6. Two coils high (add to col. 2, 3, 4, or 5; header sized for both coils) | | 7. Three coils high (add to col. 2, 3, 4, or 5; header sized for all three coils) | | Pipe size, in | Maximum gpm to each coil |
|---|---|---|---|---|---|---|---|---|---|---|---|---|---|---|---|---|
| | Total, $ | $/gpm | Total, $ | $/gpm | Total, $ | $/gpm | Total, $ | $/gpm | Total, $ | $/gpm | Total, $ | $/gpm | Total, $ | $/gpm | | |
| | 615 | 153.75 | 755 | 188.75 | 865 | 216.25 | — | — | — | — | — | — | — | — | $\frac{3}{4}$ | 4 |
| | 700 | 87.50 | 860 | 107.50 | 995 | 124.35 | 820 | 102.50 | 845 | 52.80 | — | — | — | — | 1 | 8 |
| | 790 | 49.45 | 905 | 56.55 | 1060 | 66.25 | 920 | 57.50 | 950 | 59.40 | 1300 | 40.62 | — | — | $1\frac{1}{4}$ | 16 |
| | 845 | 33.85 | 1045 | 41.80 | 1210 | 48.40 | 980 | 39.20 | 1015 | 40.60 | 1504 | 30.08 | 2134 | 28.45 | $1\frac{1}{2}$ | 25 |
| | 1050 | 21.00 | 1450 | 29.00 | 1650 | 33.00 | 1225 | 24.50 | 1265 | 25.30 | 1690 | 16.90 | 2511 | 25.10 | 2 | 50 |
| | 1870 | 23.40 | 2200 | 27.50 | 2785 | 34.80 | 2395 | 29.95 | 2585 | 32.30 | 2705 | 16.90 | 4244 | 17.70 | $2\frac{1}{2}$ | 80 |
| | 2230 | 15.90 | 2615 | 18.70 | 3250 | 23.20 | 2775 | 19.80 | — | — | 3107 | 11.10 | 5029 | 12.00 | 3 | 140 |

**Exhibit 7.4**
**COOLING, HEATING, AND HVAC EQUIPMENT AND PIPING COST ESTIMATING**
**SUMMARY SHEET**

|  | Total cost, $ |
|---|---|
| Fan coil assemblies | _____ |
| Unit ventilator assemblies | _____ |
| Transmission mains:<br>Chilled water<br>Heating water (enter on Heating Summary Sheet) | _____ |
| Perimeter distribution piping mains:<br>Chilled water<br>Heating water (enter on Heating Summary Sheet)<br>Drain pan piping (see Plumbing) | _____ |
| HV units assemblies | _____ |
| AC units assemblies | _____ |
| HV multizone assemblies | _____ |
| HVAC multizone assemblies:<br>Two-deck<br>Three-deck<br>Double-duct | _____<br>_____<br>_____ |
| Cooling coil assemblies* | _____ |
| Chilled water coil piping assemblies* | _____ |
| Chilled water coil flexible connectors* | _____ |
| Chilled water coil pumps* | _____ |
| Chilled water primary-secondary bridles* | _____ |
| Miscellaneous_____ | |
| _____ | _____ |
| Cooling portion estimate, this sheet | _____ |

* Enter heating components on Heating Summary Sheet.

# 8

# Packaged Heating, Cooling, and HVAC Units

## Furnace-Type Equipment

### Furnaces

We priced electric, gas, and oil-fired furnaces from the Lennox Co., Dallas, Texas. All models were designed by the manufacturer for add-on cooling and priced with motors large enough to accommodate cooling coils. All furnaces were priced with electric ignition. Figure 8.1 shows the schematic assemblies used in pricing gas and oil furnaces, including the piping to same. The oil tank is not included in the oil furnace price; see Special Systems (Chapter 14) for oil tank pricing. Curves on Figure 8.2 show the cost per MBH furnace output.

Add thermostat costs to these costs: heating only, $75; heating and cooling, $100.

Installed costs for options include:

| | |
|---|---|
| Flue damper | $150 |
| Electronic air cleaner | $310 |
| Humidifier, spray type | $300 |
| LP gas conversion | $ 15 |

No power wiring is included in any assembly cost. Combustion air is assumed to come from the space; no cost was assigned.

### Condensing units and coils

The assembly used for pricing the cooling side applicable to the furnace is shown in Figure 8.3a, with cost in dollars per ton shown on Figure

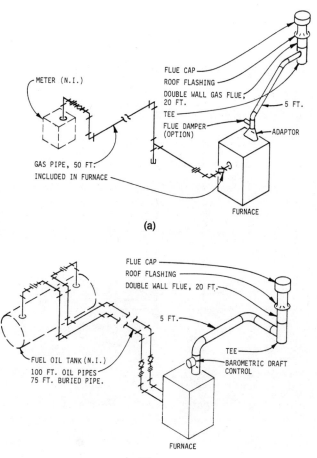

(a)

(b)

**Figure 8.1**  Furnace assembly, (a) gas fired; (b) oil fired.

8.3b. The units were all chosen for 230-V single-phase power at 105°F ambient, 80° room temperature, and 450 cfm/ton to match the cubic-foot-per-minute rating chosen for the furnaces. Follow Lennox recommendations for the maximum tons that you may add on to any given furnace.

Power wiring is not included.

### Unit heaters

Prices for gas unit heaters were arrived at using the assembly shown in Figure 8.4a. Costs are shown on Figure 8.4b. Heaters were chosen with electric ignition.

Add $75 for cost of wall-mounted thermostat. Also add cost of the gas

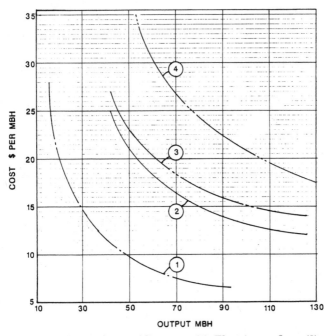

Figure 8.2  Furnace assemblies cost. (1) Electric up flow; (2) gas-fired up flow; (3) gas-fired horizontal; (4) oil-fired up flow.

main required to connect to the heater gas line. No power wiring is included.

## Packaged Cooling Units (Rooftop)

### 2- to 5-Ton Units

Lennox units were priced and Figure 8.5a shows the costs of 2- to 5-ton units. Using 105° ambient, 80° dry-bulb (DB), 67° wet-bulb (WB) temperatures, costs were plotted at actual capacities listed. Curves show costs of the basic unit with manual, 0 to 25% minimum outside air damper. A second curve shows the cost of the unit with an economizer, including outside air–return air package, exhaust damper package, and duct enclosure. A third curve shows the cost of a heat pump unit with economizer included. All costs included the price of motors and starters, 2-in throwaway filters, a roof-mounting frame, and a cooling thermostat.

Units were chosen for 230-V single-phase power, but no power wiring was included.

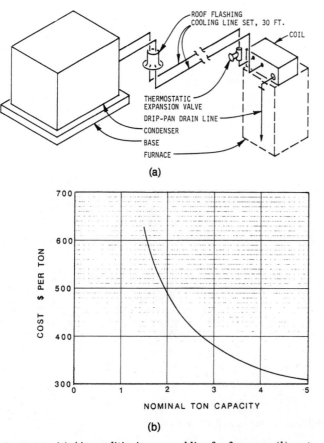

(a)

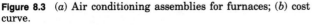

(b)

**Figure 8.3**  (a) Air conditioning assemblies for furnaces; (b) cost curve.

Figure 8.5b shows the cost in dollars per kilowatt for adding electric heat to the 2- to 5-ton units. Costs are for the largest size heater available for each unit size. Follow Lennox Co. recommendations for the maximum size heater applicable to each cooling unit. Cost of heating and cooling thermostat is included.

### 7.5- to 20-ton units

Figure 8.6a shows the costs of 7.5- to 20-ton cooling units. All conditions were the same as for 2- to 5-ton units except unit costs were plotted for nominal tonnage because it was very close to the actual tonnage, and units were chosen for three phase power.

Figure 8.6b shows the cost for adding electric heat to the 7.5- to 20-ton cooling units. All conditions were the same as for 2- to 5-ton units.

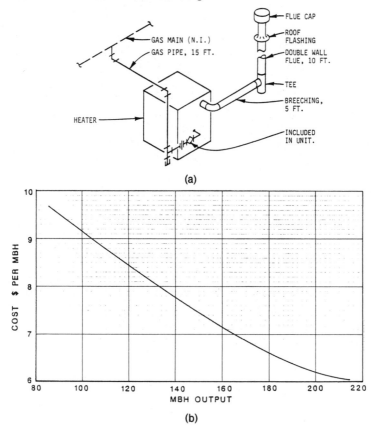

(a)

(b)

**Figure 8.4**  (*a*) Gas-fired unit heater assembly; (*b*) cost curve.

## Packaged Heating & Cooling Units (Rooftop)

### 2- to 5-ton units (Lennox)

Figure 8.7 shows the assembly used for pricing these units. Curves on Figure 8.7*a* show the cost of the basic unit with manual, 0 to 25% minimum outside air damper. A second curve shows the cost of the unit with economizer, including outside air/return air package, exhaust damper package, and duct enclosure. Costs included the price of motors and starters, 2-in throwaway filters, a roof-mounting frame, cooling section and gas heating section, and heating and cooling thermostat. Using 105° ambient, 80° DB, 67° WB, and 450 cfm/ton, the cooling capacities were very close to nominal tonnage ratings, thus no correction was made in plotting the curves.

**Figure 8.5**  (opposite page) (*a*) Packaged cooling units, 2 to 5 tons. (1) Basic unit with minimum outside air; (2) unit with economizer; (3) heat pump unit with economizer. (*b*) cost curve for electric heat installed in 2- to 5-ton cooling or heat pump units.

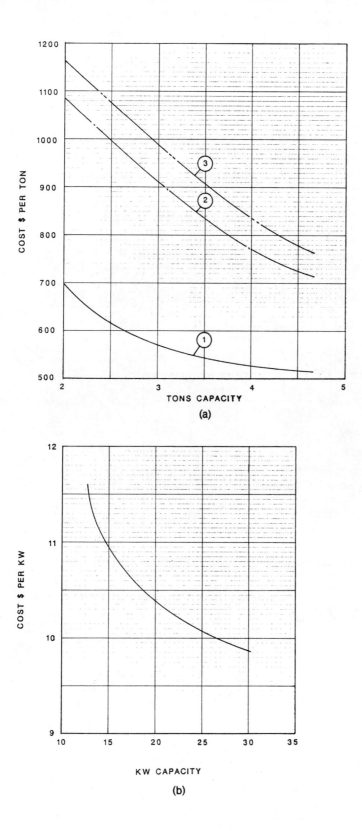

(a)

(b)

KW CAPACITY

139

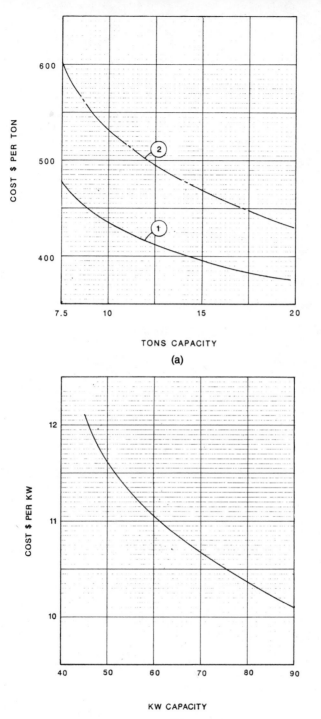

**Figure 8.6** (*a*) Cost curves for packaged cooling units, 7.5-20 tons. (1) Cooling unit with minimum outside air; (2) Cooling unit with economizer and gravity exhaust. (*b*) Cost curve for electric heat installed in 7.5-20 ton cooling units.

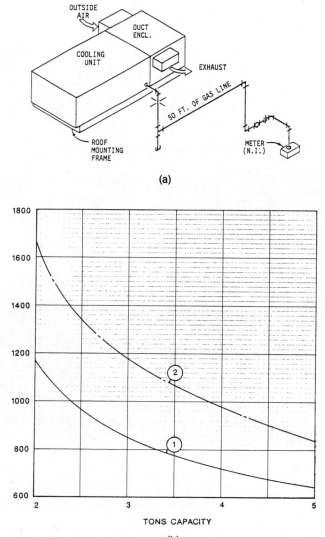

**Figure 8.7** (a) Packaged heating and cooling units, 2 to 5 tons; (b) cost curve: (1) basic unit with minimum outside air; (2) unit with economizer (power saver).

Units were chosen for 230-V single-phase power but no power wiring was included.

### 7.5- to 20-ton units (Lennox)

These units have a slightly different configuration than shown for 2- to 5-ton units; however, Figure 8.7a is adequate to show the assembly used

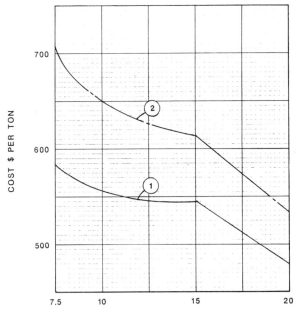

TONS CAPACITY

**Figure 8.8**  Cost curves for packaged heating and cooling units, 7.5 to 20 tons. (1) Basic unit with minimum outside air; (2) unit with economizer and gravity exhaust.

for pricing these units. Figure 8.8 shows the costs. All conditions were the same as for 2- to 5-ton units except units were chosen for three-phase power.

### 20- to 60-ton units (Trane)

Again, these units have a different configuration than shown for 2- to 5-ton units, but Figure 8.7a is adequate to show the assembly used for pricing these units.

Curves on Figure 8.9 show the cost of the basic unit with manual, 0 to 25% minimum outside air damper. A second curve shows the cost of the unit with economizer, including outside air–return air dampers and barometric relief dampers.

Costs include the price of motors and starters, 2-in thick (25%) efficiency filters, a roof-mounting frame, cooling section and gas heating section, and a remote heating and cooling thermostat.

Using 105° ambient, 80° DB, 67° WB, and 450 cfm/ton at 1.5-in TSP, the cooling capacities were slightly less than the nominal tonnage. Costs were plotted for the actual tonnage at these conditions.

Units were chosen for 230-V, three-phase power, but no power wiring was included.

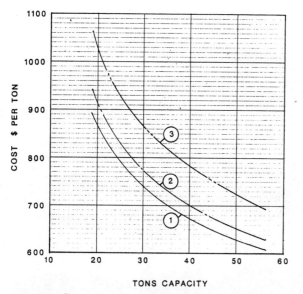

TONS CAPACITY

**Figure 8.9**  Cost curves for packaged heating and cooling units, 20 to 60 tons. (1) Basic unit with minimum outside air; (2) unit with economizer and barometric relief; (3) VAV unit with economizer and barometric relief.

### 20- to 60-ton VAV units (Trane)

All conditions were the same for VAV units except we added 1-in SP for a TSP of $2\frac{1}{2}$ when sizing the supply fan motor. Inlet vanes and additional controls for VAV operations were included in the cost. The gas heating section is used for morning warmup and night heating.

**Optional equipment available for 20- to 60-ton units.**  Remote monitoring panel with thermostat, night setback, clock, and morning warmup; cost is $500.

### Evaporative Coolers

All costs shown on Figure 8.10 are for commercial installations and include

25 ft of $\frac{1}{2}$-in copper water line with shutoff valve in the building, hose bibb at the unit for flushing and manual drain valve in the building

1-in overflow line connected to 1-in drain line in the building downstream of a drain line valve and 25 ft of 1-in copper drain line

Roof flashings for piping

Bleed-off pet cock to overflow

Winter cover

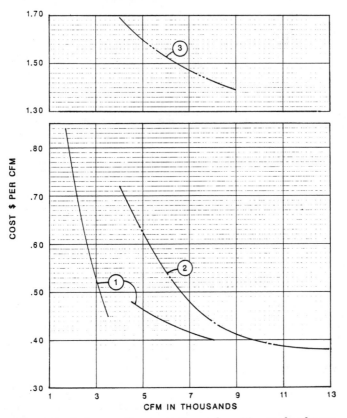

**Figure 8.10**   Cost curves for evaporative coolers. (1) wetted pad type; (2) slinger type; (3) slinger type with gas-fired heater for 100% makeup air.

For heating makeup air units we included 50 ft of gas line to heater, including five elbows, two unions, gas cock, flashing, tee, nipples, and cap.

### Wetted-pad-type cooler

Units were chosen at 0.3-in SP with two-speed motors. The air quantities at this static pressure are reduced from the nominal cubic-foot-per-minute rating of the units. Units were priced from Arvin Industries Inc., Phoenix, Arizona.

Costs included the cooler, motor and drive, wet pad, labor, and two-speed off switch and wiring. Starter is to be furnished to others for installation. Units are to be mounted on a base furnished by others.

Units were chosen for 115-V single-phase current through 3500 cfm on curve 1 and 230/60/3 current above that.

Exhibit 8.1
**PACKAGED COOLING, HEATING, AND HVAC EQUIPMENT COST ESTIMATING
SUMMARY SHEET**

|  | Total cost, $ |
|---|---|
| Furnace assemblies | _____ |
| Options | _____ |
| Thermostats (heating or heating and cooling) | _____ |
|  |  |
| Air conditioning assemblies for furnaces | _____ |
|  |  |
| Unit heater assemblies | _____ |
| Thermostats | _____ |
|  |  |
| Packaged cooling unit assemblies |  |
| 2–5 tons with minimum outside air | _____ |
| 2–5 tons with economizer | _____ |
| 2- to 5-ton heat pumps | _____ |
| Electric heat option | _____ |
|  |  |
| 7.5–20 tons with minimum outside air | _____ |
| 7.5–20 tons with economizer | _____ |
| Electric heat option | _____ |
|  |  |
| Packaged heating and cooling unit assemblies |  |
| 2–5 tons with minimum outside air | _____ |
| 2–5 tons with economizer | _____ |
|  |  |
| 7.5–20 tons with minimum outside air | _____ |
| 7.5–20 tons with economizer | _____ |
|  |  |
| 20–60 tons with minimum outside air | _____ |
| 20–60 tons with economizer | _____ |
| 20- to 60-ton VAV units | _____ |
| Options | _____ |
|  |  |
| Evaporative cooler assemblies |  |
| Wetted-pad type | _____ |
| Slinger type | _____ |
| Makeup air with heating | _____ |
|  |  |
| Miscellaneous_____ | _____ |
| Portion of estimate, this sheet | _____ |

**Slinger-type cooler**

Units were chosen at 0.5-in SP assuming they would be used on larger
installations with more extensive ductwork than employed with the
wetted-pad-type cooler. Unit cubic-foot-per-minute rating was chosen
to keep the efficiency between 73 and 76%.

Costs included the cooler with prefilters and dual cooling filters, motor and drive, cool-off-vent switch, and wiring. Starter to be furnished to others for installation. Units mounted on a base furnished by others. Units were priced from Alton Manufacturing Co., Dallas, Texas. Units were chosen for 230/60/3 current.

### Makeup air units

Units were Alton coolers with gas-fired duct heaters. Three sizes were priced, on the smallest size, 4,000 cfm, the heater output was 300,000 Btu/hr; at 6,000 cfm, 600,000 Btu/hr; and at 9,000 cfm, 900,000 Btu/hr.

A heating thermostat with heating-cooling-vent switch with wiring was added and included in the cost shown on curve 3, Figure 8.10.

### Summary

Exhibit 8.1 is the cost estimating summary sheet for packaged cooling, heating, and HVAC equipment

Chapter

# 9

# Air Distribution Equipment

Figure 9.1, from AMCA Publication 201, shows their recommendations for fan selection. We tried to select each fan on the right-hand side of the recommended selection range. For forward-curve (FC) fans this approximates the 60% wide-open cfm (WOCFM) curve, for other fans it approximates the 70% WOCFM curve.

Curves on Figure 9.2 show the costs of

Centrifugal fans

Axial fans

Cabinet fans

Vaneaxial fans

All prices were from the Trane Co. and include motor and starter for 460-V power. Starter is furnished but not installed.

### Fans

#### Centrifugal fans

Curve 1 on Figure 9.2 shows the cost of Class I, double-width (DW), air-foil fans at 2 ½-in TSP. Curve 2 shows the cost of Class II, DW, air-foil fans at 5 ½-in TSP.

Fans were arrangement 3 and included the fan, fan and motor isolation base and springs, motor slide rail base, fixed-pitch sheaves, V-belt drive and guard, motor and starter, and 4-in concrete pad.

For VAV operation, the added cost of inlet vanes and actuators is $1,000 for each fan.

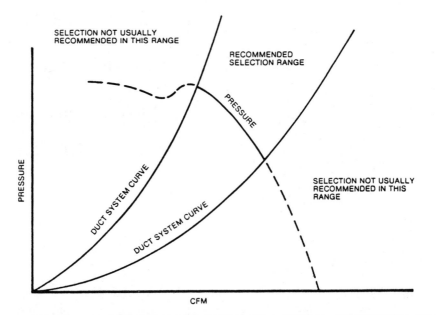

RECOMMENDED PERFORMANCE RANGE OF A TYPICAL CENTRIFUGAL FAN

**Figure 9.1**   Fan selection graph. (*From AMCA Publication 201.*)

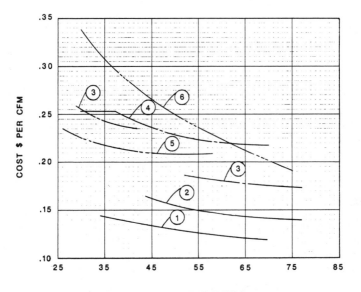

CFM IN THOUSANDS

**Figure 9.2**   Fans costs. (1) AFDW, Class I, 2 ½ in TSP; (2) AFDW, Class II, 5 ½ in TSP; (3) axial, Class I, 2 ½ in TSP; (4) axial, Class II, 5 ½ in TSP; (5) cabinet, 2½ in TSP; (6) vaneaxial, 4½ in TSP.

## Axial fans

Curve 3 on Figure 9.2 shows the cost of Class I fans at 2 $\frac{1}{2}$ in TSP. Curve 4 shows the cost of Class II fans at 5 $\frac{1}{2}$ in TSP.

Fans were arrangement 1 except for the higher cubic-foot-per-minute range in curve 3. These are arrangement 9 fans, arrangement 1 not available. Arrangement 1 is more expensive than arrangement 9. All of the accessories included for centrifugal fans were included in these costs.

For VAV operation, the added cost of inlet vanes and actuators is the same as for centrifugal fans.

## Cabinet fans

Curve 5 on Figure 9.2 shows the cost of cabinet fans at 2 $\frac{1}{2}$ in TSP. All of the accessories included for centrifugal and axial fans were included in these costs.

For VAV operation, the added cost of inlet vanes and actuators is the same as for centrifugal fans.

## Vaneaxial fans

Curve 6 on Figure 9.2 shows the cost of vaneaxial fans at 4 $\frac{1}{2}$ in TSP, based on a plenum inlet and ducted outlet with diffuser. Fans are variable-pitch fans, 1,750 rpm. Costs include the fan, diffuser with duct flange, motor and starter, isolation base and springs on 4-in concrete pad. Costs also include $1,040 for VAV control. Starter furnished but not installed.

## Exhaust Fans

The costs of several types of fans are shown on Figure 9.3. Where a choice of fans for a given duty was available, we selected the different fans on the same criteria. The fan selection graph in Figure 9.1 and the same selection procedure used for selecting fans shown in Figure 9.2 were used in selecting these fans.

Fans were priced with 230/60/1 current up to $\frac{1}{3}$ hp. For $\frac{1}{3}$ hp and up we used 230/60/3 current to price the motors and starters. Fans were all V-belt drive with variable-pitch sheaves. No controls were included.

Fans were priced from the Trane Co., Loren Cook Co., and Penn Ventilator Co.

### Dome-type roof exhaust fans and low-contour exhaust fans

Both series were selected at a maximum tip speed of 3,500 ft/min and a maximum sone value of 9.0. Fans were selected at $\frac{1}{2}$ in TSP.

NOTE: CURVE 2 & 3 CAN BE EXTRAPOLATED TO $.22/CFM AT 14,000 CFM.

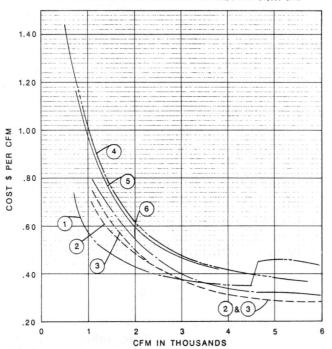

**Figure 9.3** Exhaust fans costs. (1) Dome-type roof exhaust fan, $\frac{1}{2}$ in TSP; (2) dome-type upblast, 1 in TSP; (3) dome-type upblast kitchen hood exhaust fan, 1-in TSP; (4) low-contour roof exhaust fan, $\frac{1}{2}$ in TSP; (5) utility fan, $\frac{1}{2}$ in TSP; (6) utility fan, 2 in TSP.

Included in the costs shown on Figure 9.3 were the fan, motor and drive, backdraft damper, and labor. Twelve-inch-high sound curb and starter furnished for installation by others.

As a general rule, dome-type *wall* exhaust fans cost an additional $.05/cfm more than the dome type for roof mounting.

**Utility fans**

Utility fans were generally selected on the 65% WOCFM line. Curves on Figure 9.3 show the cost per cubic foot per minute for fans at $\frac{1}{2}$ and at 2 in TSP.

Costs include fan, motor and drive, motor hood, scroll damper volume control, backdraft dampers, isolation base with rubber in shear isolators, and labor. Starter furnished for installation by others. Roof curb by others.

For Heresite coating add $150 to the price of the fan.

## Dome-type upblast exhaust fans

Curve 2 on Figure 9.3 shows the cost of fans suitable for laboratory hood exhaust. They were chosen at 1 in TSP and include the cost of the fan, motor and drive, backdraft dampers, labor, and Eisen-Heiss or epoxy coating. Twelve-inch-high curb and starter furnished for installation by others.

Curve 3 on Figure 9.3 shows the cost of fans suitable for restaurant hood exhaust, UL listed to meet National Fire Protection Association (NFPA) 96 requirements. Fans were chosen at 1 in TSP and include the cost of the fan, motor and drive, grease trough, and labor. Starter and UL listed 18-in-high curb furnished for installation by others.

Fans were chosen only to meet the SP requirement and no regard was given to sound criteria. This probably accounts for their prices being cheaper in the larger sizes than the down-blast configuration shown in the curves 1 and 4.

## Propeller fans

A curve is shown on Figure 9.4 for small direct-drive fans selected at 1,040 rpm and $\frac{1}{4}$ in TSP. Prices include the fan and motor, installation, and backdraft damper. They were selected for 230/60/1 current. Starters furnished to others for installation.

A curve is also drawn for larger belt-drive fans with steel blades selected at $\frac{1}{2}$ in TSP. We sometimes use these for return-air fans so no backdraft dampers were included in the price. Prices did include the fan, motor and drive, front and rear fan guard, and installation. Starter furnished to others for installation.

## Residential fans

Fans were all priced installed on the lower floor of a two-story residence with duct through the roof. Wiring for the fan unit is not included. Table 9.1 shows these costs.

## Built-up filter banks

Costs include frames and filters with installation costs in built-up banks using 24 × 24 in filters with velocities ranging from 450 to 525 ft/min for system sizes of 36,000 to 80,000 cfm. Costs did not vary by size in this range. Prices were from Air Filter Sales & Service Co., Denver, Colorado. Table 9.2 shows these costs.

## Prefabricated plenums

Sound-attenuating fan plenums may be priced on a square foot basis. The estimator must make a rough layout of the fan rooms with all

NOTE: CURVE 2 BELT DRIVE MAY BE EXTRAPOLATED TO .7¢/ CFM AT 44,500 CFM

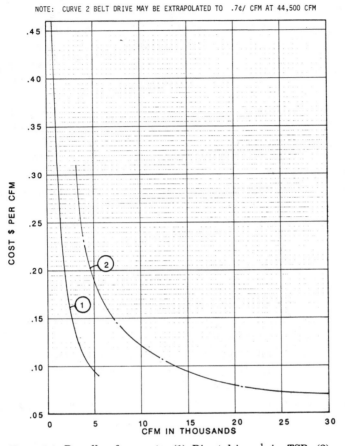

**Figure 9.4** Propeller fans costs. (1) Direct-drive, $\frac{1}{4}$ in TSP; (2) belt-drive, $\frac{1}{2}$ in TSP.

### TABLE 9.1 Residential Fans

| Fan type | Cost |
| --- | --- |
| Toilet exhaust fan, 110 cfm, with 15-min timer wall switch and roof cap | $ 160 |
| Deluxe toilet exhaust fan, 460 cfm, with 15-min timer wall switch and roof cap | $ 280 |
| Kitchen range hood, 30–36 in, 250 cfm, with light and fan switches, fan, and roof cap | $ 360 |
| Kitchen range hood, stainless steel or hammered brass with light and fan switches, fan, and roof cap. | $ 410 |
| Oven hood, 27-in stainless steel with fan, fan switch, and roof cap | $ 385 |

TABLE 9.2    Built-up Filter Banks, 36,000 to 80,000 CFM Load

| Filter banks | Cost, $/cfm |
|---|---|
| 30% efficient filters, Farr 30/30 | .035 |
| 50% efficient filters with 2-in glass fiber prefilter, Farr NS 15, 21-in-deep bag | .055 |
| 80% efficient filters with 2-in-glass fiber prefilter, Farr NS 100, 21-in-deep bag | .06 |
| 50% efficient filters with 2-in ULOK FP 100 (<20% efficient), Farr NS 15, 21-in-deep bag | .06 |
| 80% efficient filters with 2-in ULOK FP 100 (<20% efficient), Farr NS 100, 21-in-deep bag | .065 |

walls and partial walls shown. The estimate should include the total square feet of all walls and *the ceiling*. Installed costs for base, walls, corners, ceiling, and access doors for 3- or 4-in thick steel insulated panels should run about $14.00/ft$^2$

**Sound attenuators**

Sound attenuators become more necessary as the static pressure and speed of the fans increase. For conceptual estimating purposes, we might assume that sound attenuators are required (*in both the supply and return*) on all but the smallest systems. After the job is far enough along in design for sound calculations to be made, the requirements may be dropped if calculations warrant.

Our experience shows that, generally speaking, 7-ft-long attenuators are required. Prices here are based on a face velocity of 1,225 ft/min for the IAC Company MS model silencers (medium pressure drop). Prices are given for 3-, 5-, and 7-foot-long attenuators. Rectangular attenuators are cheaper than conical attenuators even with duct transitions.

Prices are in dollars per cubic foot per minute. Be sure to use both the supply and return cubic-foot-per-minute requirement

3 ft long = $.04/cfm

5 ft long = $.06/cfm

7 ft long = $.09/cfm

For sizes with face area of 1 ft$^2$ or less increase these costs by $0.3/cfm.

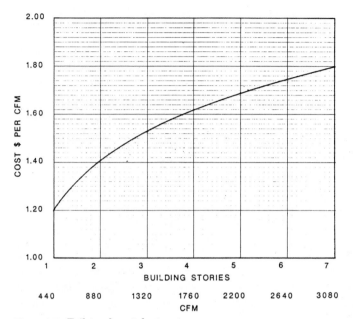

**Figure 9.5**  Toilet exhaust duct cost.

## Ductwork

The curve on Figure 9.5 shows the cost of toilet exhaust ductwork, and Figure 9.6 shows the cost of supply, return, and outside air ductwork for single-zone systems. Table 9.3 shows the costs of ductwork for larger single-zone systems and for VAV systems. If you are insulating the ductwork on the outside rather than using lined ductwork, add the appropriate costs from Chapter 12 to the ductwork costs.

Ductwork costs can vary a great deal from one project to another, depending on many factors. Some of these factors are design criteria used; duct accessories used; sound criteria employed; fan room proximity to the distribution network; average size and, consequently, the number of outlets; fire dampering required; block load versus room load diversity in a VAV system; space restrictions for the ductwork (requiring offsets, transitions, larger aspect ratio ducts, etc.); and the simplicity or complexity of the duct layout required from the design of the building. Also, whether the distribution system is end-fed or center-fed from the fan room changes the costs in a way similar to the discussion for pipe distribution in Chapter 7. Because of all of these possible variations in ductwork, the estimator's judgment is infinitely more important here than in any other area. From the systems I have studied, duct costs can vary 100% from each other or 50% from the norm. When preliminary drawings are completed it would be wise for

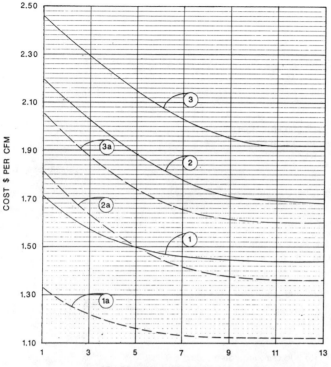

COST $ PER CFM

CFM PER AIR HANDLER, IN THOUSANDS

**Figure 9.6** Systems ducts costs for galvanized steel single-zone units. (1) Supply air duct with 1-in duct liner; (2) supply air duct plus return air duct with 1-in duct liner; (3) supply air and return air ducts plus outdoor air duct with 1-in duct liner; (1a) supply air duct, no liner; (2a) supply air duct plus return air duct, no liner; (3a) supply air and return air ducts plus outdoor air duct, no liner.

the designer to price the duct systems using the data from Table 9.4a, 9.4b, and 9.4c for a more accurate estimate than is attainable from the conceptual estimate figures.

### Conceptual estimates of systems

Supply ductwork costs shown in the figures here are costs of each duct system and include flexible connections to the fans, ductwork, fire dampers, turning vanes in square elbows, extractors at registers, twist-in fittings with volume damper for all round duct connections, flexible ducts (limited to 8 ft long by Denver code), connection to VAV boxes, and register and diffuser boxes. Add the cost of VAV boxes and diffusers to these costs.

Return ductwork costs are for projects without extensive ducted systems. For single-zone systems the costs apply for any single-story

**TABLE 9.3** **Systems Duct Costs for Single-Zone and VAV Units, 15,000 to 65,000 CFM**

| Item | Cost, $/cfm |
|------|-------------|
| System sizes 15,000–25,000 cfm, single-zone systems up to $2\frac{1}{2}$ in SP | |
| Lined supply air duct | 1.45 |
| Unlined construction | 1.13 |
| | |
| System sizes 15,000–65,000 cfm, ducts to VAV units only, average VAV system $2\frac{1}{2}$ to 6 in SP | |
| Lined supply air duct | 1.45 |
| Unlined construction | 1.13 (block load cfm) |
| | |
| Lined supply air duct, very quiet sound criteria system, $2\frac{1}{2}$–6 in SP, system sizes 15,000–65,000 cfm, ducts to VAV units only | 1.60 (block load cfm) |
| | |
| Lined supply air duct, VAV systems up to $2\frac{1}{2}$ in SP, system sizes 15,000–65,000 cfm, ducts from VAV unit to diffusers | .50 (room load cfm) |
| | |
| System sizes 15,000–65,000 cfm, return air duct | |
| Lined | .25 |
| Unlined | .20 |
| | |
| System sizes 15,000–65,000 cfm, outside air duct | |
| Lined | .25 |
| Unlined | .20 |

**TABLE 9.4a** **Ductwork Costs\***

| Size of largest duct, in | Gauge | Up to $2\frac{1}{2}$ in SP | | Up to $2\frac{1}{2}$ to 6 in SP | | 304 stainless steel duct, welded and flanged construction | 18 gauge black steel, welded and flanged construction |
|---|---|---|---|---|---|---|---|
| | | Unlined construction | Lined with 1-in, 2-lb. insulated construction | Unlined construction | Lined with 1-in, 2-lb. insulated construction | | |
| Up to 12 | 26 | .44 | .58 | .47 | .63 | 1.13 | 2.46 |
| 13–30 | 24 | .50 | .64 | .54 | .70 | 1.58 | 2.46 |
| 31–54 | 22 | .60 | .74 | .64 | .80 | 1.88 | 2.46 |
| 55–84 | 20 | .69 | .83 | .74 | .90 | 2.26 | 2.46 |
| 85 and up | 18 | .74 | .88 | .79 | .95 | 3.10 | 2.46 |

\* Cost is given in dollars per inch of surface per lineal foot of duct. Multiply this number by the sum (in inches) of the width and depth of the duct and by the length of the duct in feet to get the total cost of the duct in dollars.

TABLE 9.4b   Fittings for Galvanized Steel Ducts*

| Description | Equivalent length of duct, ft |
|---|---|
| Outlet (grille, register, diffuser) box | 3 (branch size) |
| 90° radius elbow | 3 |
| 90° reducing elbow | 5 |
| 90° elbow with turning vanes | 7 |
| 45° radius elbow | 2 |
| 45° reducing elbow | 4 |
| Transition | 3 |
| Transition with offset | 6 |
| Stub in branch take-off with extractor | 3 (branch size) |
| Branch take-off with splitter damper and 90° R ell | 6 |
| Double branch take-off with two splitters and two 90° R ells | 9 |
| Duct end cap | 1 |

* Measure all ducts through the fittings, add these lengths as extra lengths.

TABLE 9.4c   Round Ductwork and Accessories Costs

| Round size, in | Spiral wound duct, $/ft | Insulated flexible duct, $/ft* | Bell-mouth twist-in fitting, $ each* |
|---|---|---|---|
| 4 | 2.60 | 2.60 | 10 |
| 6 | 3.20 | 3.20 | 12 |
| 8 | 4.35 | 4.35 | 14 |
| 10 | 5.80 | 5.40 | 17 |
| 12 | 7.45 | 7.15 | 20 |
| 14 | 9.85 | 8.80 | 25 |
| 16 | 12.85 | 11.75 | 28 |
| 18 | 15.95 | 15.60 | 33 |
| 20 | 17.35 | 16.50 | 37 |

* Flexible duct costs are based on duct insulated with 1-in thick insulation with spring helix and inner and outer liner. Twist-in fittings costs based on bell-mouth fittings with adapter plate for insulated ducts, butterfly damper with quadrant operator, and steel strap with worm-screw tightener.

building where the return air is ducted directly from one or two return air grilles or from an elbow in the ceiling space to the air handling equipment. For multistory buildings the costs apply where the return air enters a duct in a chase through a return-air grille or elbow from the ceiling space and the air handler can be ducted rather directly to the chase. Costs include ductwork, accessories such as turning vanes and fire dampers, register boxes, and flexible connection to the air handler. *Not* included in the cost are transfer air costs for such items as grilles, openings, fire-dampered openings, etc., to transfer air from room to room or above the ceiling to the point of return air pickup.

Outside air ductwork costs are valid for air handlers located in close proximity to a wall or roof intake. The ductwork and flexible connection to the air handler is included. The cost of the intake hood or louver is *not* included and must be added to these costs.

Toilet exhaust systems shown here are for each exhaust system and include ductwork, turning vanes, fire dampers, register boxes, and flexible connection to the fan. Costs are for *unlined* ductwork. If lined ductwork is required, you may multiply the cost from Figure 9.5 by 1.3 to arrive at a conceptual cost figure. Costs on Figure 9.5 would still be viable if the cubic-foot-per-minute rating varied by 10%. Each floor includes men's and women's toilets and janitors' closet exhaust. Add the cost of registers to these costs.

### Block load vs. room load totals

For VAV systems there is a diversity between the instantaneous or block load and the maximum load per room or room load summary. This diversity varies by project. For estimating purposes in the conceptual stage of the project only a rough block load may be known. However, the VAV units; the ductwork from the VAV units to the diffusers; and the diffusers, registers, and grilles should be estimated from the room load summary. Until both the block load and a room load summary are calculated, use a diversity of 85% and divide the block load by .85 to reach a hypothetical total room load summary.

### Hoods and louvers

Hood prices with and without dampers are shown in Figure 9.7. Hoods were steel stationary hoods with $\frac{3}{4}$-in expanded metal screen, installed on curb furnished by others.

Exhaust hoods were sized at .05-in wg loss at the load being handled. Intake hoods were sized at .10-in wg loss at the load being handled.

Automatic dampers had self-lubricating bearings with seals on blade edges only and not on sides. No actuator was included.

Curves on Figure 9.7 show costs of 16-gauge galvanized steel louvers with prime coat finish and of 12-gauge extruded aluminum louvers with clear anodized finish. All louvers were 4 in deep, 45° blade angle with return bend.

Stationary louvers were sized at .10-in wg loss on intake louvers and .20-in w.g. loss on exhaust louvers. Costs include $\frac{1}{2}$-in mesh screen.

### Dampers

Table 9.5 shows the costs of several types of dampers.

**Manual opposed blade dampers.**   Dampers priced were with steel blades and frame with quadrant-type lock and handle.

**Counterbalanced dampers.**   Aluminum blade dampers in steel frame with ball bearings were the dampers priced.

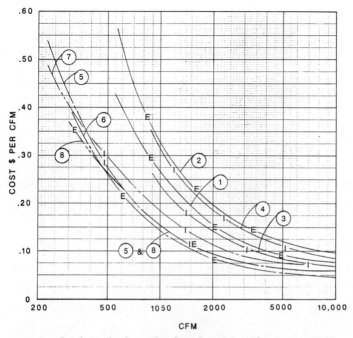

**Figure 9.7**  Intake and exhaust hoods and stationary louvers costs. (1) Exhaust hood; (2) exhaust hood with backdraft dampers; (3) intake hood; (4) intake hood with automatic dampers; (5) steel intake louver; (6) steel exhaust louver; (7) aluminum intake louver; (8) aluminum exhaust louver.

**Automatic dampers.**  Dampers priced were with steel blade and frame and seals at mating blades only. Operators were not included.

**Fire dampers.**  Dampers priced were steel shutter, type B curtain dampers. Dampers are with stacked blades completely out of the air

**TABLE 9.5  Installed Costs of Dampers**

| Size, depth by width, in | Face area, ft² | Cost of manual opposed-blade damper, $ | Cost of counter-balanced dampers, $ | Cost of fire damper in duct, $ |
|---|---|---|---|---|
| 12 × 12 | 1 | 53 | 88 | 38 |
| 12 × 24 | 2 | 63 | 94 | 42 |
| 24 × 24 | 4 | 78 | 111 | 61 |
| 24 × 36 | 6 | 91 | 145 | 80 |
| 24 × 48 | 8 | 103 | 153 | 98 |
| 36 × 36 | 9 | 124 | 165 | 107 |
| 36 × 48 | 12 | 142 | 199 | 129 |
| 36 × 72 | 18 | 230 | 240 | 168 |
| 48 × 72 | 24 | 275 | 301 | 186 |
| 48 × 96 | 32 | 321 | 357 | 234 |
| 60 × 96 | 40 | 389 | 430 | 255 |

stream. Type A dampers with blades in the air stream would run $1/ft$^2$ less. Horizontally mounted dampers would cost approximately 10% more.

### VAV control units

Curves in Figure 9.8$a$ show the installed costs of various types of VAV control units. Figure 9.8$b$ shows the costs of these units with reheat coils, including runout piping costs the same as for reheat coils in Chapter 7. Prices on the units were from the Trane Co. and were for open-end-arrangement boxes with damper motors and interior insulation. All pneumatic box prices included pressure-independent volume regulators. Electrically controlled units do not have this feature. Room

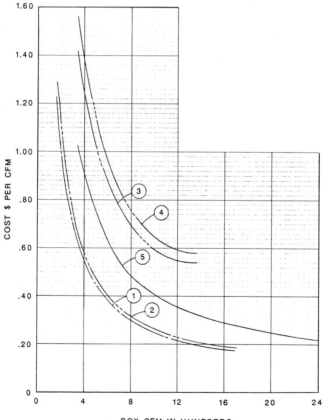

**Figure 9.8$a$**  VAV control units costs. (1) Shutoff type with pneumatic control; (2) shutoff type with electric control; (3) fan-powered type with pneumatic control; (4) fan-powered type with electric control; (5) dual duct type with pneumatic control.

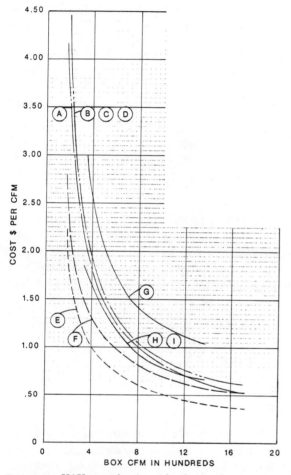

**Figure 9.8b** VAV control units with reheat coils costs. (A) Shutoff type with pneumatic control, one-row water coil; (B) shutoff type with pneumatic control, two-row water coil; (C) shutoff type with electric control, one-row water coil; (D) shutoff type with electric control, two-row water coil (add $.04/cfm to curve C); (E) shutoff type, electric coil with magnetic contactor; (F) shutoff type, electric coil with mercury contactor; (G) fan-powered type with pneumatic control, one-row water coil; (H) fan-powered type, electric coil with magnetic contactor; (I) fan-powered type, electric coil with mercury contactor (add $.10/cfm to curve H).

thermostats, night set-back, and heating changeover controls for morning warmup were not included. These must be included in the temperature control section.

For electric reheat, 208/60/3 power was assumed as the power source. Two types of contactors were priced, mercury contactors being a silent type of contactor.

Dual duct boxes were priced with summer cooling air requirement being supplied through both sets of ductwork and through both sides of the box, each side with pneumatic volume regulators.

For all VAV control units we used a nominal cubic-foot-per-minute load for each size box as follows:

| Size | 02 | 200 cfm |
|------|----|---------|
|      | 04 | 400 cfm |
|      | 08 | 800 cfm |
|      | 12 | 1,200 cfm |
|      | 20 | 2,000 cfm |

These values have been arrived at in our offices as conservative maximum values for use in areas of average sound attenuation where room sound levels are to be kept to NC 38 or below. Since each unit cannot be chosen at its maximum allowable load, we have applied a factor here of .85 to the above and assumed the average load for each of these boxes, when selected on a project, will be

| Size | 02 | 170 cfm |
|------|----|---------|
|      | 04 | 340 cfm |
|      | 08 | 680 cfm |
|      | 12 | 1,020 cfm |
|      | 20 | 1,700 cfm |

All costs were plotted at these volumes including this load through each side of the dual duct boxes.

To estimate the number of VAV units, the *maximum room load* summary should be used. Divide this load by the anticipated number of VAV units to get the average load per unit in cubic feet per minute. See ductwork, discussion under Block Load vs. Room Load Totals.

For conceptual estimating, we have found that the typical load per VAV unit in office buildings, administrative areas, hospitals, etc., is about 600 cfm per unit. For other types of buildings the estimator must make a judgment as to the number of units to use in zoning the building and enter the curves at the appropriate load per unit.

### Grilles, registers, and diffusers

Curves in Figure 9.9 show the installed cost of various grilles, registers, and diffusers. All registers and diffusers were priced with volume-control dampers and were chosen with a static pressure drop of .07-in wg.

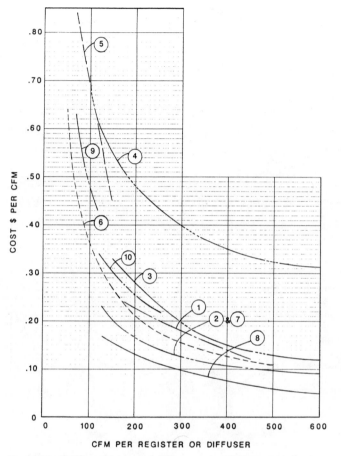

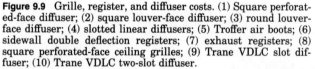

**Figure 9.9**  Grille, register, and diffuser costs. (1) Square perforated-face diffuser; (2) square louver-face diffuser; (3) round louver-face diffuser; (4) slotted linear diffusers; (5) Troffer air boots; (6) sidewall double deflection registers; (7) exhaust registers; (8) square perforated-face ceiling grilles; (9) Trane VDLC slot diffuser; (10) Trane VDLC two-slot diffuser.

Square perforated-face diffusers were priced with round neck and off-white baked-enamel finish. Square louver-face diffusers were priced with square neck and aluminum paint finish. Round louver face diffusers were priced with off-white baked-enamel finish.

The slotted linear diffuser's price included the diffuser and plenum, anodized aluminum finish, and black painted interior. Prices are for one-, two-, and four-slot diffusers and are good for 4- or 8-ft lengths.

Troffer air boot prices were one- or two-slot for 4-ft fixtures.

Sidewall registers were double deflection grilles and included volume dampers and volume extractor; grilles have off-white baked-enamel finish and aspect ratios from 2.5 to 1 up to 5 to 1.

**Exhibit 9.1**
**AIR MOVING EQUIPMENT COST ESTIMATING SUMMARY SHEET**

|  | Total cost, $ |
|---|---|
| Centrifugal fans | _____ |
| Inlet vanes and controls | _____ |
|  |  |
| Axial fans | _____ |
| Inlet vanes and controls | _____ |
|  |  |
| Cabinet fans | _____ |
| Inlet vanes and controls | _____ |
|  |  |
| Vaneaxial fans (controls included) | _____ |
|  |  |
| Dome-type roof exhaust fans | _____ |
|  |  |
| Dome-type wall exhaust fans | _____ |
|  |  |
| Utility fans | _____ |
| Coating | _____ |
|  |  |
| Laboratory upblast exhaust fans | _____ |
|  |  |
| Kitchen hood upblast exhaust fans | _____ |
|  |  |
| Propeller fans (small DD) | _____ |
|  |  |
| Propeller fans (large BD) | _____ |
|  |  |
| Residential exhaust fans |  |
| Hoods | _____ |
| Toilets | _____ |
|  |  |
| Miscellaneous_____ | _____ |
|  |  |
| Portion of estimate, this sheet | _____ |

Exhaust registers were sized at approximately 300 ft/min entrance velocity; grilles have off-white baked-enamel finish.

Return grille prices are for perforated-face ceiling grilles with off-white baked-enamel finish in metal frame, sizes 12 × 12 in through 24 × 24 in. They were sized with static pressure losses of .03 in. Other types of sidewall grilles have similar prices. Lay-in 24 × 24 in perforated panels with no frame run from $.10/cfm at 150 cfm each to $.04/cfm at 500 cfm each.

Curves 9 and 10 on Figure 9.9 are for Trane VDLC diffusers for T-bar ceilings with plenums included. Prices for these units in 4-ft sections used as return air grilles with no plenums are approximately the same as for curve 8, perforated-face returns.

Exhibit 9.2
AIR DISTRIBUTION EQUIPMENT COST ESTIMATING SUMMARY SHEET

| | Total cost, $ |
|---|---|
| Built-up filter banks | _____ |
| Prefabricated plenums | _____ |
| Sound attenuators | _____ |
| Ductwork | |
| Single-zone SA, RA, OA, all systems | _____ |
| Toilet exhaust, all systems | _____ |
| VAV SA to boxes, RA and OA, all systems | _____ |
| VAV SA from boxes, all systems | _____ |
| Hoods | |
| Intake | _____ |
| Exhaust | _____ |
| Louvers | |
| Intake | _____ |
| Exhaust | _____ |
| Dampers | |
| Manual opposed blade | _____ |
| Counterbalanced | _____ |
| Automatic | _____ |
| Fire | _____ |
| VAV control units | |
| Shutoff type | _____ |
| Fan-powered type | _____ |
| Dual-duct type | _____ |
| Shutoff with reheat | _____ |
| Fan-powered with reheat | _____ |
| Grilles, registers, diffusers | _____ |
| Miscellaneous_____ | _____ |
| Portion of estimate, this sheet | _____ |

Residential-quality grilles, registers, and diffusers for low-cost projects cost less than the ones priced here for commercial work.

All of these prices are for medium-sized jobs of $250,000 to $500,000. Very large jobs might see discounts of up to 10% on the material prices or 5% on the total installed price.

To estimate the number of outlets, the *maximum room load* summary should be used. Divide this load by the anticipated number of outlets to get the average load per outlet in cubic feet per minute. See ductwork discussion under Block Load vs. Room Load Totals.

For conceptual estimating, we have found that the typical load per outlet in office buildings, administration areas, hospitals, etc. is about 300 cfm. Clinics have smaller rooms and run 200 cfm per outlet. For other types of buildings the estimator must make a judgment as to the value to use.

### Summary

Exhibits 9.1 and 9.2 show the cost estimating summary sheets for air moving equipment and air distribution equipment, respectively.

# 10

# Controls and Instrumentation

The control systems for projects researched were done by independent firms doing only control systems on mechanical projects.

## The Conceptual Estimate

We have compared the cost of controls to the cost of the heating and cooling portions of the mechanical contracts over the past years. These portions are represented by Chapter 7, Heating and Cooling, and Chapter 9, Air Distribution Equipment, in this book. Packaged heating, cooling, and HVAC units, described in Chapter 8, have self-contained control systems and are not included. Once you have arrived at a cost from these two chapters, you can use Figure 10.1, which shows the cost of the control contract as a percentage of these combined costs. Apply the percentage to these combined costs before assigning the contractor's profit to them. Curve 1 shows the control costs for primary and secondary schools. Curves 2 and 3 show the lower and upper limits of all other types of projects researched. Costs represented by these curves are based on past projects whose controls were primarily pneumatic wtih combination pneumatic-electric and electric components. Costs included the air compressor station, air dryers, temperature controls, interlocks, smoke and freeze control, and local enclosed control panels at each control center. A central display panel was part of each project. No central control, reset, recording, or printout was included.

CURVES 2 & 3 MAY BE EXTRAPOLATED TO 14%
& 16% AT $2,500,000 COST OF (A)

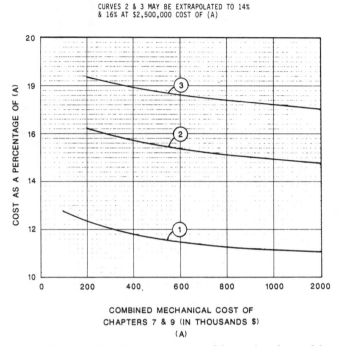

COMBINED MECHANICAL COST OF
CHAPTERS 7 & 9 (IN THOUSANDS $)
(A)

**Figure 10.1**   Control costs as a percentage of the combined cost of the mechanical work in Chapters 7 and 9 before assigning the contractor's profit. (1) Primary and secondary schools; (2) lower limit, all other types; (3) upper limit, all other types.

### More detailed estimate

After the preliminary drawings stage or the design development stage is completed, the estimator may want to make a more detailed estimate of the control costs. We have prepared a list of costs of control components plus a "sequence of control" with the costs of same for the equipment covered in Chapters 7 and 9. When using the sequence of control, the estimator must pick and choose the costs applicable to the situation.

While the controls may be accomplished by other than a pneumatic system with pneumatic-electric and electric components, these costs should give a reasonable estimate of the eventual project cost. Wiring and tubing runs were estimated assuming power and main air lines within 25 ft. Components to be included in the estimate are:

The compressed air station, including a dryer

The main air line, approximate length in feet

Each unit operation sequence-of-operation cost, plus the valve material cost (Table 10.1), plus the damper cost, which includes damper material cost and damper operator material and installation cost (Figure 10.2) and tubing runs as appropriate

Smoke and freeze controls

To these costs, you must add the cost of any central monitoring, control, or automation system, costs of which are not included here. Last, we have a control and instrumentation cost summary sheet, Exhibit 10.1, to use with the more detailed estimate. This exhibit appears at the end of this chapter.

Add the control contractor's profit to these figures to get the total subcontract price to the mechanical contractor.

*Compressed Air Station:* Costs include the air compressor, tank, intake air filter and silencer, PRVs, air dryer and isolation base set on a 4-in concrete pad.

| SIZE: | COST, $ |
|---|---|
| $\frac{1}{2}$-hp, 15-gal tank, simplex | 2,700 |
| 1-hp, 30-gal tank, simplex | 3,000 |
| 1 $\frac{1}{2}$-hp, 30-gal tank, simplex | 3,000 |
| 2-hp, 60-gal tank, simplex | 3,150 |
| 3-hp, 60-gal tank, simplex | 3,300 |
| 1-hp, 60-gal tank, duplex | 4,200 |
| 1 $\frac{1}{2}$-hp, 60-gal tank, duplex | 4,200 |
| 2-hp, 60-gal tank, duplex | 4,450 |
| 3-hp, 60-gal tank, duplex | 4,750 |

*Main Air Line*: For copper tubing, use a cost per lineal foot of    4.00

**TABLE 10.1    Pneumatic Control Valve Prices***

| | Valve size, in | | | | | | | | | | |
|---|---|---|---|---|---|---|---|---|---|---|---|
| Valve Type | $\frac{1}{2}$ | $\frac{3}{4}$ | 1 | $1\frac{1}{4}$ | $1\frac{1}{2}$ | 2 | $2\frac{1}{2}$ | 3 | 4 | 6 | 8 |
| Two-way[†] | 85 | 100 | 120 | 130 | 145 | 230 | 565 | 740 | 990 | 1,800 | 2,300 |
| Three-way[†] | 100 | 120 | 135 | 185 | 200 | 350 | 695 | 910 | 1,160 | 2,000 | 2,600 |
| One valve with two butterflies and operator | — | — | — | — | — | — | 457 | 500 | 600 | 1,100 | 1,500 |

* Prices are material prices only; the installation is accomplished by the mechanical contractor.
† With positioner, add $100.

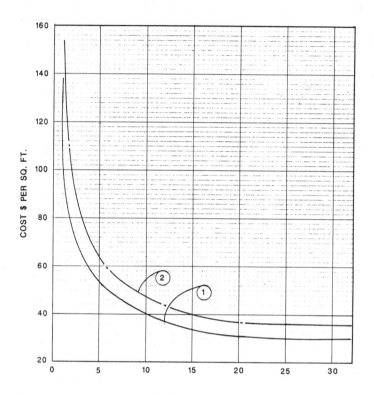

DAMPER SIZE IN SQ. FT.

**Figure 10.2**  Control damper costs. (1) Seals on blade edges only; (2) seals on blade edges plus the sides of the damper frame.

COST, $

*Pneumatic Tubing*: One-quarter-inch copper tubing runs can be estimated at the same price as plastic tubing run in conduit or in metal trough. Cost per lineal foot of    2.00

*Boiler Control*: Boiler water temperature and steam pressure    675
is maintained constant from controls furnished with the boiler. A temperature sensor, sensing outdoor air temperature, transmits a signal to a receiver/controller which allows the main gas valve to open and the burner to operate from the boiler controls when the outdoor air temperature is below 65°F. The controls cause the main gas valve to close and the boiler burner to shut off above 70°F outdoor air temperature.

*Combustion Air Preheater Control*: Face and bypass dampers,    0
damper motor, and discharge air thermostat are furnished with the Wing Co. heaters.

Connect control air to the above controls and calibrate to    175
maintain discharge air at 55°F.

A flow switch in the water line leaving the coils causes the    175
face dampers to close if hot-water flow fails.

COST, $

*Heat Exchanger (Steam-to-Fluid) Control:* A temperature sensor, sensing fluid temperature leaving the heat exchanger, transmits a signal to a receiver/controller which modulates one or two steam valve(s) to:

1. Maintain a constant temperature of the leaving fluid          400

or

2. Reset the temperature of the leaving fluid inversely with          675
   outdoor air temperature as signaled by an outdoor air
   temperature sensor

*Add cost of steam valve(s)*

When load current is applied to the pump handling the          375
liquid leaving the heat exchanger, the controls allow the
steam valve(s) to modulate. When load current is broken,
the steam valve(s) close.

*Heat Exchanger (Water-to-Fluid) Control:* A temperature sensor, sensing fluid temperature leaving the heat exchanger, transmits a signal to a receiver/controller which:

1. Cycles the heat exchanger heating water circulating          525
   pump to maintain a constant temperature of the leaving
   fluid.

or

2. Modulates a valve in the heating supply to reset the          675
   temperature of the leaving fluid inversely with outdoor
   air temperature as signaled by an outdoor air temperature sensor.

When load current is applied to the pump handling the          375
liquid leaving the heat exchanger, the controls allow the
heating water circulating pump to run and the heating water
three-way mixing valve to modulate. When load current is
broken, the heating water valve closes to the heat exchanger
and the heating water circulating pump is off.

*Add cost of mixing valve*

*Scheduled Supply Fluid Temperature Control:* A tempera-          675 for 1,
ture sensor that senses supply fluid temperature transmits a          500 each for
signal to a receiver/controller, which modulates a three-way          2 or more in
mixing valve to reset the temperature of the supply fluid          same room
inversely with outdoor air temperature as signaled by an
outdoor air temperature sensor.

One fluid temperature sensor, receiver/controller, and mixing valve for each of the following systems:

Baseboard radiation
Radiant ceiling panels
Coils
Snow melting
Other

COST, $

*Add cost of mixing valve(s)*

Pump Control

HEATING PUMPS: A temperature sensor, sensing outdoor air    450 for 1
temperature, transmits a signal to a receiver/controller
which through electrical interlock, allows each of the following pumps to run below 65°F outdoor air temperature and
stops each pump automatically above 70°F outdoor air temperature.

Main heating pump(s)                                        100 each
Heat exchanger supply fluid pump(s)                         pump for 2
Baseboard radiation pump(s)                                 or more
Radiant ceiling panels pump
Coil pumps
Other

COOLING PUMPS: Control for the primary chilled water        0
pump(s) and condenser water pump(s) is under Chiller Plant
Control.

A temperature sensor, sensing outdoor air temperature,      450 for 1
transmits a signal to a receiver/controller which through
electrical interlock allows each of the following pumps to run
above 55°F outdoor air temperature and stops each pump
automatically below 50°F outdoor air temperature.

Coil pumps                                                  100 each
Other                                                       pump for 2
                                                            or more

For systems with no antifreeze solution employed in cli-     250 for 1
mates where freezing may occur, start the following pumps
and allow them to run below 35°F outdoor temperature.

Coil Pumps                                                  100 each
Other                                                       pump for 2
                                                            or more

*Hot-Water Unit Heater and Cabinet Heater Control*: A 24-V    200
room thermostat cycles the heater fan through a 24-V relay
to maintain its setting.

*Radiation (Baseboard, Radiant Panels, Convectors) Control*:   200
A room thermostat modulates a two-way control valve in the
radiation supply line to maintain its setting.

*Add cost of temperature control valve(s)*

*Domestic Hot-Water Control*: A temperature sensor that       400
senses tank or heater-leaving water temperature transmits a
signal to a receiver/controller which modulates a two-way
control valve in the heating water line to maintain a constant domestic water temperature.

*Chiller Plant Control with Cooling Tower*: This sequence to
take place only when any time controller is on DAY cycle in
the AUTO position or on DAY manual position.

A temperature sensor, sensing outdoor air temperature,        500
transmits a signal to a receiver/controller which starts the

COST, $

primary chilled water pump and applies load current to the
cooling tower fan when the outdoor air temperature rises to
55°F. When the outdoor air temperature falls 5° below this,
the controls break the load current to the pump and cooling
tower fan.

After the cycle is started, a flow switch, located in the            350
chilled water leaving the evaporator, will apply load current
to the condenser water pump, allowing it to start when
chilled water flow is proven. A second flow switch, located in
the water line leaving the condenser, will apply load current
to the chiller compressor, allowing it to start from its controls
when flow is proven.

When the cycle is shut down by any of the above, a               250
time-delay relay will allow the chilled water and condenser
water pumps to run for an adjustable period of time. (In order
for the compressor to go through its pump-down cycle.)

When there are more than one each chilled water and         750 for 1
condenser water pumps, the controls are arranged to allow       add'l. pump
any of the pumps to fulfill the sequence of control.              for each
                                                                 system

A temperature sensor, sensing the water temperature          350
leaving the chiller, transmits a signal to a receiver/controller
which maintains a constant chilled water temperature.
These controls are furnished with the chiller. Connect and
calibrate these controls.

OUTDOOR COOLING TOWER: A temperature sensor, sensing         500
the water temperature entering the condenser, transmits a
signal to a receiver/controller which modulates two linked
butterfly valves to bypass the cooling tower to maintain a
constant water temperature to the condenser.

An immersion-type remote bulb thermostat, located in the     400
tower basin, cycles the tower fan when the water temperature falls to 5° below the desired condenser water temperature.

*Add cost of butterfly valves*

INDOOR COOLING TOWER: A temperature sensor, sensing the      450
water temperature entering the condenser, transmits a signal to a receiver/controller which modulates opposed blade
dampers in the tower intake ductwork to maintain a constant
water temperature to the condenser.

An end switch on the damper motor will interrupt load        500
current to the cooling tower fan when the dampers are closed.
When the dampers are 25% open and after a 5-min time
delay from closing, a position switch will apply load current
to the tower fan, allowing it to run.

*Add cost of automatic dampers and operators*

*Cooling Tower Freeze Protection Control*: Cooling tower basin    0

COST, $

heater(s) and controls are furnished as part of the cooling tower.

Heat tape and a contactor on each of the cold water makeup, tower drain, and tower overflow lines are furnished by the mechanical contractor. Provide a contact thermostat on each of these three lines to control the temperature at 40°. Also provide a liquid level probe in the tower basin which will break the current to all heat tapes, and to the sump heater(s) when the tower is drained.          650

Liquid level probes in the tower basin operate a two-way fill valve to maintain the water level. When a manual tower drain switch is activated, the tower fill valve closes and drain valves at the base of this line, the tower supply and return lines, and the tower drain line will open to drain the portions of these lines located outdoors.          450

*Add cost of temperature control valves*

*Chiller Plant Control, Air-Cooled*: This sequence to take place only when any time controller is on DAY cycle in the AUTO position or on DAY manual position.

A temperature sensor, sensing outdoor air temperature, transmits a signal to a receiver/controller which starts the primary chilled water pump when the outdoor air temperature rises to 55°F. When the outdoor air temperature falls 5° below this, the controls break the load current to the pump.          400

After the cycle is started, a flow switch, located in the chilled water leaving the evaporator, will apply load current to the chiller compressor allowing it to start from its controls.          175

When the cycle is shut down by any of the above, a time-delay relay will allow the chilled water pump to run for an adjustable period of time. (In order for the compressor to go through its pump-down cycle.)          175

When there is more than one chilled water pump, the controls are arranged to allow any of the pumps to fulfill the sequence of control.          375 for 1 additional pump

A temperature sensor, sensing the water temperature leaving the chiller, transmits a signal to a receiver/controller which maintains a constant chilled water temperature. These controls are furnished with the chiller. Connect and calibrate these controls.          350

CONDENSING FOR PACKAGED AIR-COOLED CHILLER/ CONDENSER: Condensing temperature is maintained by controls furnished with the chiller package.          0

CONDENSING FOR CHILLER PLANT, CONDENSERLESS WITH REMOTE AIR-COOLED CONDENSERS: Interlock the condenser fan(s) with a 120-V relay off of the load side of the chiller compressor so that the fan(s) may run when load current is applied to the compressor. Relay is furnished with the chiller.          325

COST, $

*Freeze Detection Thermostat Control*: Each air supply system    300
taking outdoor air and which has water coils or steam coils
has an electric freeze-detection thermostat with 20 ft of
flexible sensing bulb, manual reset, and an isolated set of
contacts to wire to the control alarm system. When the
temperature falls below the set point, the thermostat stops
its system supply fan.

*Smoke Detection Control (NFPA Requirements)*: In systems
of over 2,000 cfm capacity, smoke detectors approved for duct
installation are installed in

The main supply duct on the downstream side of the filters    350*
to automatically stop the fan, and

The return air stream, prior to exhausting from the build-    350*
ing or being diluted by outside air, to automatically stop the fan

EXCEPTION NO. 1: The smoke detector in the return air stream
may be omitted in systems of less than 15,000 cfm capacity.

EXCEPTION NO. 2: Both detectors may be omitted provided
that the system is less than 15,000 cfm capacity, the entire
system is within the space served, and such space is protected
by an area smoke detection system.

*HV Unit Control*: The HV unit control may also be used for    0
unit ventilator control with one night clock for each grouping
of unit ventilators. Return air and outdoor air dampers are
provided with the HV unit. Relief dampers are provided
under another section.

A control panel provided includes:    2,000
    Outdoor air temperature sensor receiver/controller
    A 7-day time clock with 10-hr reserve spring and skip-
a-day feature.
    DAY-NIGHT-AUTO switch
    Pilot lights for each of the following: DAY, NIGHT, AUTO
    With the switch in the DAY position, the DAY light is on
and the unit provides a DAY cycle operation.
    With the switch in the NIGHT position, the NIGHT light is
on, and the building remains on night setback.
    With the switch in the AUTO position, automatic opera-
tion with changeover from night setback to day temperature
is provided as set by the time clock and the pilot lights
indicate where in the cycle the unit is operating.
    Systems operate as follows:
    DAY cycle: The DAY cycle starts 1 hr before the building is
open to allow for warmup. The HV unit fan runs continuously.
    A duct thermostat, located in the return air, keeps the    350

---

* Either photoelectric or product-of-combustion type.

COST, $

return air damper open and the outdoor air damper closed until the return air temperature rises to 70°F.

A room thermostat, through a low-limit temperature sensor that senses fan discharge air temperature, transmits a signal to a receiver/controller which modulates the return air and outdoor air dampers in sequence with a two-way or three-way hot-water-coil control valve to maintain room thermostat setting. Outdoor air dampers modulate to 100% open position above 65°F outdoor air temperature.    900

*Add cost of control valve or valve included with unit ventilator*

A mixed-air low-limit thermostat limits the return air and outdoor air damper positions to keep the mixed air from going below 45°F. The mixed-air low-limit thermostat overrides all other controls to modulate the outdoor air dampers to 100% shutoff if mixed-air temperature falls below its setting. (Not required in climates where freezing does not occur.)    350

The outdoor air damper closes and the return air damper opens whenever the HV unit fan stops.    200

NIGHT cycle: During the NIGHT cycle, the HV unit fan is off. The outdoor air dampers are closed and the return air dampers are open.    0

or

During the NIGHT cycle, a "night" thermostat cycles the HV unit fan on a wide differential to maintain a reduced temperature. The outdoor air damper is closed and the return air damper is open.    300

The heating coil control valve is positioned to full flow from the heat source through the coil whenever the HV unit fan goes off and whenever it is on the NIGHT cycle. A flow switch, located in the coil discharge piping, sounds an alarm, turns on a red warning light at the control panel, and stops the HV unit fan when a no-flow condition is sensed. (Cannot be used with two-way control valve.) A manual alarm silencer is provided. Alarm and fan interlock is locked out when the pump is turned off from the outside air sensor, which normally controls the operation. (Not required in climates where freezing does not occur.)    700

*HV Multizone Unit Control (No Return Air Fan):* Return air and outdoor air dampers are provided with the HV unit. Relief dampers are provided under another section.    0

A control panel provided includes:    2,000
    Outdoor air temperature sensor receiver/controller
    A 7-day time clock with 10-hr reserve spring and skip-a-day feature
    DAY-NIGHT-AUTO switch

COST, $

Pilot lights for each of the following: DAY, NIGHT, AUTO
With the switch in the DAY position, the DAY light is on
and the unit provides a DAY cycle operation
With the switch in the NIGHT position, the NIGHT light is
on, and the building remains on night setback
With the switch in the AUTO position, automatic operation with changeover from night setback to day temperatures
is provided as set by the time clock and the pilot lights
indicate where in the cycle the unit is operating

Systems operate as follows:
DAY cycle: The DAY cycle starts 1 hr before the building is
open to allow for warmup. The unit fan runs continuously.
A duct thermostat, located in the return air, keeps the            350
return air damper open and the outdoor air damper closed
until the return air temperature rises to 70°F.
Each zone thermostat modulates its zone hot and bypass        400 each
deck dampers to maintain its setting.
Hot deck: A load selector system and hot deck controller          700
modulate a two-way or three-way valve on the heating coil to
maintain the lowest hot deck temperature required to satisfy
the zone calling for the most heating.
Bypass deck: The high-pressure output of the load selector        375
modulates the outdoor air and return air dampers through an
"economizer cycle" to provide the highest bypass deck temperature required to satisfy the zone calling for the most
cooling. Outdoor air dampers modulate to a 100% open
position above 75°F outdoor air temperature.

*Add cost of control valve*

A mixed-air low-limit thermostat limits the return air and        350
outdoor air dampers to keep the mixed air from going below
45°F. The mixed-air low-limit thermostat overrides all other
controls to modulate the outdoor air dampers to 100% shutoff
if mixed-air temperature falls below its setting. (Not required in climates where freezing does not occur.)
The outdoor air damper closes and the return air damper          200
opens whenever the unit fan stops.
NIGHT cycle: During the NIGHT cycle, the unit fan is off. The       0
outdoor air dampers are closed and the return air dampers
are open.

or

During the NIGHT cycle, a "night" thermostat cycles the           300
unit fan on a wide differential to maintain 60°F minimum
temperature. The outdoor air damper is closed and the return
air damper is open.
The heating coil control valve is positioned to full flow from    700
heat source through the coil whenever the unit fan goes off
and whenever it is on the NIGHT cycle. A flow switch, located

COST, $

in the heating coil discharge piping, sounds an alarm, turns on a red warning light at the control panel, and stops the unit fan when a no-flow condition is sensed. (Cannot be used with two-way control valve.) A manual alarm silencer is provided. Alarm and fan interlock is locked out when the pumps are shut off from outside air sensor, which normally controls the operation. (Not required in climates where freezing does not occur.)

*HVAC Unit Control (No Return Air Fan):* The HVAC unit    0
control may also be used for heating and cooling unit venti-lator control with one night clock for each grouping of unit ventilators. Return air and outdoor air dampers are provided with the HVAC unit. Relief dampers are provided under another section.

A control panel provided includes:    2000
Outdoor air temperature sensor receiver/controller
A 7-day time clock with 10-hr reserve spring and skip-a-day feature
DAY-NIGHT-AUTO switch
Pilot lights for each of the following: DAY, NIGHT, AUTO.
With the switch in the DAY position, the DAY light is on and the unit provides a DAY cycle operation
With the switch in the NIGHT position, the NIGHT light is on, and the building remains on night setback
With the switch in the AUTO position, automatic opera-tion with changeover from night setback to day temperatures is provided as set by the time clock and the pilot lights indicate where in the cycle the unit is operating.

Systems operate as follows:
DAY cycle: The DAY cycle starts 1 hr before the building is open to allow for warmup or cool-down. The unit fan runs continuously.

A duct thermostat, located in the return air duct, keeps the    350
return air damper open and the outdoor air damper closed until the return air temperature rises to 70°F.

A room thermostat, through a low-limit temperature sen-    1,050
sor that senses fan discharge air temperature, transmits a signal to a receiver/controller which modulates the return air and outdoor air dampers in sequence with a two-way or three-way hot-water coil and a two-way or three-way chilled water coil control valve to maintain room thermostat setting. Outdoor air dampers modulate to a minimum open position above 75°F outdoor air temperature.

*Add cost of control valves*

A mixed-air low-limit thermostat limits the return air and    350
outdoor air dampers to keep the mixed air from going below 45°F. The mixed-air low-limit thermostat overrides all other controls to modulate the outside air dampers to 100% shutoff

COST, $

if mixed-air temperature falls below its setting. (Not required in climates where freezing does not occur.)

The outdoor air damper closes and the return air damper    200
opens whenever the unit fan stops.

NIGHT cycle: During the NIGHT cycle, the unit fan is off. The    0
outdoor air dampers are closed and the return air dampers
are open.

or

During the NIGHT cycle, a "night" thermostat cycles the    300
unit fan on a wide differential to maintain 60°F minimum
temperature. The outdoor air damper is closed and the return
air damper is open.

The heating coil control valve is positioned to full flow from    700
heat source through the coil whenever the unit fan goes off
and also on the NIGHT cycle. A flow switch, located in the
heating coil discharge piping, sounds an alarm, turns on a
red warning light at the control panel, and stops the unit fan
when a no-flow condition is sensed. (Cannot be used with a
two-way control valve.) A manual alarm silencer is provided.
Alarm and fan interlock is locked out when the pump is shut
off from outside air sensor, which normally controls the operation. (Not required in climates where freezing does not occur.)

The chilled water cooling coil control valve is positioned to    200
full flow from the chiller through the coil whenever the unit
fan is off if the outdoor temperature is below 35°F. (Not
required in climates where freezing does not occur or if
system is filled with antifreeze solution.)

*AC Unit Control (No Return Air Fan):* The AC unit control    0
may also be used for cooling only unit ventilator control with
one night clock for each grouping of unit ventilators. It was
assumed that any unit used in climates subject to freezing
would also have a heating coil and that this unit would not be
used if freezing could occur. Return air and outdoor air
dampers are provided with the AC unit. Relief dampers are
provided under another section.

A control panel provided includes:    2,000
Outdoor air temperature sensor receiver/controller
A 7-day time clock with 10-hr reserve spring and skip-
a-day feature
DAY-NIGHT-AUTO switch
Pilot lights for each of the following: DAY, NIGHT, AUTO
With the switch in the DAY position, the DAY light is on
and the unit provides a DAY cycle operation
With the switch in the NIGHT position, the NIGHT light is
on, and the building remains on night setback
With the switch in the AUTO position, automatic operation with changeover from night setback to day temperatures
is

COST, $

provided as set by the time clock, and the pilot lights indicate where in the cycle the unit is operating.

Systems operate as follows:

DAY cycle: The DAY cycle starts 1 hr before the building is open to allow for cool-down. The unit fan runs continuously.

A room thermostat, through a low-limit temperature sensor that senses fan discharge air temperature, transmits a signal to a receiver/controller which modulates the return air and outdoor air dampers in sequence with a two-way or three-way chilled water coil control valve to maintain room thermostat setting. Outdoor air dampers modulate to a minimum open position above 75°F outdoor air temperature.    900

*Add cost of control valves*

The outdoor air damper closes and the return air damper opens whenever the unit fan stops.    200

NIGHT cycle: During the NIGHT cycle, the unit fan is off. The outdoor air dampers are closed and the return air dampers are open.    0

*HVAC, Two-Deck Multizone Unit Control (No Return Air Fan):* Return air and outdoor air dampers are provided with the HVAC unit. Relief dampers are provided under another section.

A control panel provided includes:    2,000
    Outdoor air temperature sensor receiver/controller
    A 7-day time clock with 10-hr reserve spring and skip-a-day feature
    DAY-NIGHT-AUTO switch.
    Pilot lights for each of the following: DAY, NIGHT, AUTO
    With the switch in the DAY position, the DAY light is on and the unit provides a DAY cycle operation
    With the switch in the NIGHT position, the NIGHT light is on, and the building remains on night setback
    With the switch in the AUTO position, automatic operation with changeover from night setback to day temperatures is provided as set by the time clock and the pilot lights indicate where in the cycle the unit is operating.

Systems operate as follows:

DAY cycle: The DAY cycle starts 1 hr before the building is open to allow for warmup or cool down. The unit fan runs continuously.

A duct thermostat, located in the return air, keeps the return air damper open and the outdoor air damper closed until the return air temperature reaches 70°F.    350

Each zone thermostat modulates its zone hot and cold deck dampers to maintain its setting.    400 each

Hot deck: A load selector system and hot deck controller modulate a two-way or three-way valve on the heating coil to    700

COST, $

maintain the lowest hot deck temperature required to satisfy
the zone calling for the most heating.

Cold deck: The high-pressure output of the load selector     500
modulates a two-way or three-way valve on the chilled water
coil in sequence with the outdoor air and return air dampers
through an "economizer cycle" to provide the highest cold
deck temperature required to satisfy the zone calling for the
most cooling. Outdoor air dampers modulate to a minimum
open position above 75°F outdoor air temperature.

*Add cost of control valves*

A mixed-air low-limit thermostat limits the return air and     350
outdoor air dampers to keep the mixed air from going below
45°F. The mixed-air low-limit thermostat overrides all other
controls to modulate the outside air dampers to 100% shutoff
if mixed-air temperature falls below its setting. (Not required in climates where freezing does not occur.)

The outdoor air damper closes and the return air damper     200
opens whenever the unit fan stops under any condition.

NIGHT cycle: During the NIGHT cycle, the unit fan is off. The     0
outdoor air dampers are closed and the return air dampers
are open.

or

During the NIGHT cycle, a "night" thermostat cycles the     300
unit fan on a wide differential to maintain a reduced temperature. The outdoor air damper is closed and the return air
damper is open.

The heating coil control valve shall be positioned to full     200
flow from heat source through the coil whenever the unit fan
goes off and also on the NIGHT cycle.

The chilled water cooling coil control valve is positioned to     200
full flow from the chiller through the coil whenever the fan is
off if the outdoor temperature is below 35°F. (Not required in
climates where freezing does not occur or if system is filled
with antifreeze solution.)

Flow switches, located in both the heating coil discharge     1,000
piping and the cooling coil discharge piping, sound an alarm,
turn on a red warning light at the control panel, and stop the
unit fan when a no-flow condition is sensed by either flow
switch. (Cannot be used with two-way control valves.) A
manual alarm silencer is provided. Alarm and fan interlock
is locked out when the pumps are shut off from outdoor air
sensor, which normally controls the operation. (Not required
in climates where freezing does not occur.)

*VAV Fan System Control (with Return Air Fan):* Return
air and outdoor air dampers are provided with the VAV unit.
Exhaust air dampers are provided under this section.

or

COST, $

Return air, outdoor air, and exhaust air dampers are provided under this section.

*Add the cost of dampers*

A control panel provided includes:                                    2,800
Outdoor air temperature sensor receiver/controller
An optimal start programmer to create an optimal start time for any outside air temperature
DAY-NIGHT-AUTO switch
Pilot lights for each of the following: DAY, NIGHT, AUTO
With the switch in the DAY position, the DAY light is on and the unit provides a DAY cycle operation.

With the switch in the NIGHT position, the NIGHT light is on, and the building remains on night setback

With the switch in the AUTO position, automatic operation with changeover from night setback to day temperatures through a MORNING WARMUP cycle is provided as set by the programmer and the pilot lights indicate where in the cycle the unit is operating.

Systems operate as follows:
DAY cycle: When the DAY cycle starts, the supply air fan      300
runs continuously. The return air fan is interlocked to run when the supply air fan runs.

A duct thermostat, sensing supply air fan discharge air      1,200
temperature, transmits a signal to a receiver/controller which modulates the return air, outdoor air, and exhaust air dampers in sequence with a two-way or three-way hot-water coil control valve and a two-way or three-way chilled water control valve to maintain thermostat setting of 55°F. Outdoor air and exhaust air dampers modulate to a minimum open position above 65°F outdoor air temperature.

*Add cost of control valves*

A mixed-air low-limit thermostat limits the return air,       350
exhaust air, and outdoor air damper positions to keep the mixed air from going below 45°F. The mixed-air low-limit thermostat overrides all other controls to modulate the outdoor air and exhaust air dampers to 100% shutoff if mixed-air temperature falls below its setting. (Not required in climates where freezing does not occur.)

The outdoor air and the exhaust air dampers close and the     300
return air damper opens whenever the supply air fan stops.

NIGHT cycle: *Systems with shutoff VAV units or fan-          0
powered VAV units or systems in climates not requiring night heating.* Supply air and return air fans are off. The outdoor and exhaust air dampers are closed and the return air dampers are open.

*Systems with VAV units with reheat coils.* A "night"         300
thermostat cycles the unit fan on a wide differential to maintain 60°F minimum temperature. The outdoor air and

COST, $

exhaust air dampers are closed and the return air damper is open.

The heating coil pump is on. The heating coil control valve is modulated by a temperature sensor in the supply air fan discharge and a panel-mounted receiver/controller set at 75°F (adjustable).     1,000

See VAV units control for their operation.     0

MORNING WARMUP cycle: *Systems with shutoff VAV units or*     1,500
*with VAV units with reheat coils.* The supply air and return air fans are on. The outdoor air and exhaust air dampers are closed and the return air damper is open. The heating coil pump is on. The chilled water coil control valve is to be positioned for recirculated flow through the coil. The heating coil control valve is modulated by a temperature sensor in the supply air fan discharge and a panel-mounted receiver/ controller set at 75°F (adjustable).

See VAV units control for their operation.     0

The heating coil control valve is positioned to full flow from     700
heat source through the coil whenever the unit fan goes off and also on the NIGHT cycle. A flow switch, located in the coil discharge piping, sounds an alarm, turns on a red warning light at the control panel, and stops the supply air fan when a no-flow condition is sensed. (Cannot be used with two-way control valve.) A manual alarm silencer is provided. Alarm and fan interlock is locked out when the pump is shut off from outdoor air sensor, which normally controls the operation. (Not required in climates where freezing does not occur.)

The chilled water cooling coil control valve is positioned to     200
full flow from the chiller through the coil whenever the fan is off if the outdoor temperature is below 35°F. (Not required in climates where freezing does not occur or if system is filled with antifreeze solution.)

Static pressure sensors transmit signals to a receiver/     2,000
controller which modulates the supply air fan and return air fan variable-inlet vanes (furnished with the fans) to maintain a constant static pressure. The sensor is located in the supply duct two-thirds of the way to the last VAV control unit in a section of ductwork with minimum turbulence. Sensing device is a multiple point, nonpulsating static pressure sensing section with self-averaging manifold. Provide gauges visible through the control panel face to read the cfm of each fan.

or

*Static Pressure Control Portion (three stage):* The static     1,800
pressure for supply fan control is sensed two-thirds of the way to the last VAV control unit in a section of ductwork with minimum turbulence. The sensor modulates vane or pitch control through a receiver controller and high-limit static control in the fan discharge. The controllers are panel-

COST, $

mounted. Provide gauge visible through the control panel
face to read the cfm of the fan.

Return fan static is sensed in the return/exhaust plenum.    1,500
The sensor modulates variable speed, variable pitch, or inlet
vanes to maintain setpoint through a receiver/controller and
low-limit static sensor located in the return duct ahead of the
return fan. Controllers are panel-mounted. Provide gauge
visible through the control panel face to read the cfm of the fan.

The building static is sensed in the occupied space and      900
referenced to outside. The sensor modulates the exhaust
damper through a receiver/controller. Controller is consid-
ered panel-mounted.

*VAV Fan System Control (with Return Air Fan) No Heating
Coil.* It was assumed that any unit used in climates subject to
freezing would also have a heating coil and that this unit
would not be used if freezing could occur.

Return air and outdoor air dampers are provided with the
VAV unit. Exhaust air dampers are provided under this
section.

or

Return air, outdoor air, and exhaust air dampers are
provided under this section.

*Add the cost of dampers*

A control panel provided includes:                          2,000
    Outdoor air temperature sensor receiver/controller
    A 7-day time clock with 10-hr reserve spring and skip-
a-day feature
    DAY-NIGHT-AUTO switch
    Pilot lights for each of the following: DAY, NIGHT, AUTO.
    With the switch in the DAY position, the DAY light is on
and the unit provides a DAY cycle operation
    With the switch in the NIGHT position, the NIGHT light is
on, and the building remains on night setback
    With the switch in the AUTO position, automatic opera-
tion with changeover from night to day temperatures is
provided as set by the time clock, and the pilot lights indicate
where in the cycle the unit is operating.

Systems operate as follows:
    DAY cycle: When the DAY cycle starts, the supply air fan
runs continuously. The return air fan is interlocked to run
when the supply air fan runs.

A duct thermostat, sensing supply air fan discharge air      1,050
temperature, transmits a signal to a receiver/controller
which modulates the return air, outdoor air, and exhaust air
dampers in sequence with a two-way or three-way chilled
water control valve to maintain thermostat setting of 55°F.
Outdoor air and exhaust air dampers modulate a minimum
open position above 65°F outdoor air temperature.

COST, $

*Add cost of control valve*

The outdoor air and the exhaust air dampers close and the return air damper opens whenever the supply air fan stops under any condition.   300

NIGHT cycle: During the NIGHT cycle, the supply air and return air fans are off. The outdoor air and exhaust air dampers are closed and the return air dampers are open.   0

Static pressure control, see VAV fan system control.

*Add appropriate cost*

*Dual-Duct VAV System, Heating Side Fan Control:* It was assumed this unit has no outdoor air or exhaust air and that no return air dampers are required.

A manually operated "summer-winter" switch will set the operating mode.   200

A control panel includes:   2,000
Outdoor air temperature sensor receiver/controller
An optimal start programmer to create an optimal start time for any outdoor air temperature
DAY-NIGHT-AUTO switch
Pilot lights for each of the following: DAY, NIGHT, AUTO
With the switch in the DAY position, the DAY light is on and the unit provides a DAY cycle operation
With the switch in the NIGHT position, the NIGHT light is on, and the building remains on night setback
With the switch in the AUTO position, automatic operation with changeover from night setback to day temperatures through a MORNING WARMUP cycle is provided as set by the programmer and the pilot lights indicate where in the cycle the unit is operating.

Systems operate as follows:

DAY cycle: When the DAY cycle starts, the supply air fan runs continuously. The dual-duct VAV system cooling side supply fan will be interlocked to run when this fan runs.   300

A duct thermostat, sensing supply air fan discharge air temperature, transmits a signal to a receiver/controller which modulates in sequence a two-way or three-way hot-water coil control valve and a two-way or three-way chilled water control valve to maintain a scheduled air temperature, scheduled inversely with outdoor temperature.   900

*Add cost of control valves*

The heating side control of the VAV units served by this fan system must be served by a separate pneumatic air line (in addition to the main pneumatic air line) which permits this system to change over from VAV to constant volume when the manually operated switch is changed from "winter" to "summer" setting.   4.00/ft

COST, $

NIGHT cycle: A "night" thermostat cycles the unit fan on a    300
wide differential to maintain a reduced temperature.

See VAV units' control for their operation.    0

MORNING WARMUP cycle: Control is the same as for DAY    300
cycle except chiller start control is locked out.

See VAV units' control for their operation.    0

Static pressure control, see VAV fan system control.

*Add appropriate cost*

Fan Coils: One control panel with night clock and DAY-    800 each
NIGHT-AUTO cycle for each grouping of fan coils provided.

COOLING FAN COILS: DAY cycle: The DAY cycle starts 1 hr    0
before the building is open to allow for cool-down. The fan coil
units' fans run continually.

The fan coil automatic outdoor air damper opens when fan    100
power is restored. Damper closes when fan stops.

A room thermostat modulates the chilled water coil control    200 for 1 fan
valve on the fan coil unit(s) under its control to maintain    coil on one
room thermostat setting.    thermostat,
     100 for each
     additional

A low-limit thermostat with 48-in flexible sensor will close    200
the outdoor air damper on sensing freezing temperature.

*Valves included with fan coil*

*Damper and operator included with fan coil*

NIGHT cycle: During the NIGHT cycle, the fan coil fans are    0
off. The fan coil automatic outdoor air damper is closed.

HEATING-COOLING FAN COILS: DAY cycle: The DAY cycle starts    0
1 hr before building is open to allow for warmup or cool-down.
The fan coil units' fans run continuously.

The fan coil automatic outdoor air damper opens when fan    100
power is restored. Damper closes when fan stops.

A room thermostat modulates the heating coil control    350 for 1 fan
valve in sequence with the cooling coil control valve on the    coil on one
fan coil unit(s) under its control.    thermostat,
     150 for each
The thermostat modulates the heating coil valve when the    additional
room temperature is 72°F and below (set point adjustable).
The cooling coil valve is in the closed position. Between 72
and 75°F (adjustable) room temperature, the thermostat has
the heating coil valve closed and the cooling coil valve closed.
When the room temperature is above 75°F (adjustable), the
cooling coil valve is modulated to maintain room temperature and the heating coil valve remains closed.

*Valves included with fan coil*

*Damper and operator included with fan coil*

NIGHT cycle: During the NIGHT cycle, a "night" thermostat    300
cycles its zone fan coil fans on a wide differential to maintain
a reduced temperature.

COST, $

The fan coil automatic outdoor air damper is closed.    0

*Shut-Off VAV Unit Control:* Connect air to the volume   25/unit
regulator furnished with the VAV unit.

DAY cycle: A room thermostat modulates the damper oper-   200 for 1
ator on the VAV unit(s) under its control. Damper is modu-   VAV unit,
lated open on demand for cooling to maintain the thermostat   100 for each
setting.   additional
  unit on one
  thermostat

NIGHT cycle: Same control for VAV units as DAY cycle.   0
A "night" thermostat modulates radiation and radiant   500
heating control valves to maintain a reduced temperature.

MORNING WARMUP cycle: The dampers for all VAV units   100/unit
are driven open to their assigned maximum load to allow
constant-temperature air to be circulated to the space.

*VAV Unit with Reheat Coil (or with Radiation or Radiant*   25/unit
*Heating) Control:* Connect air to the volume regulator furnished with the VAV unit.

DAY cycle: This sequence is for each VAV unit and valve
under the control of one thermostat.

A room thermostat modulates a normally open heating   350 for 1
valve when the room temperature is 72°F and below (set   VAV unit,
point is adjustable). The VAV unit damper operator is at its   150 for each
minimum position setting. When the room temperature is   additional
between 72 and 75°F (adjustable), the heating coil valve is   unit on one
closed and the VAV unit damper operator is at its minimum   thermostat
setting. When the room temperature is above 75°F (adjustable), the VAV unit damper operator modulates to maintain
room temperature and the heating valve remains closed.

NIGHT cycle: Reheat coil control valves and VAV unit   500
dampers are under the control of their DAY thermostat. A
"night" thermostat modulates radiation and radiant heating
control valves to maintain a reduced temperature.

MORNING WARMUP cycle: The dampers for all VAV units   100/unit
are driven open to their assigned maximum load to allow a
constant-temperature air to be circulated to the space. Reheat, radiation, and radiant heating control valves are under
the control of their DAY thermostat.

*Add cost of valve(s)*

*Fan-Powered VAV Unit, Parallel Configuration, with Reheat*   25/unit
*Coil:* Connect air to the volume regulator furnished with the
VAV unit.

DAY cycle: This sequence is for each VAV unit and valve
under the control of one thermostat.

When the room temperature is above 75°F (adjustable), a   350 for 1
room thermostat modulates the VAV unit damper operator   VAV unit,
to maintain room temperature. The VAV unit fan is off.   150 for each
When the room temperature is between 72 and 75°F (adjust-   additional
able), the VAV unit damper is closed and the VAV unit fan   unit on one
is started and circulates return air from the ceiling plenum to   thermostat

COST, $

the space. When the room temperature falls to 72°F (adjustable) and below, the thermostat modulates the reheat coil valve while the VAV unit fan continues to run and the VAV unit damper remains closed.

NIGHT cycle: A night thermostat cycles the fan on its zone VAV unit(s) on a wide differential to maintain a reduced temperature.    300

MORNING WARMUP cycle: Control is restored to the VAV unit's DAY thermostat.    0

*Add the cost of valve(s)*

*Dual-Duct VAV Unit Control:* Connect air to the two volume regulators furnished with the VAV unit.    25/unit

A manually operated "summer-winter" switch will initiate the mode; see VAV fan system control.    0

A room thermostat modulates the damper operators on the VAV unit(s) under its control. The following sequences are for summer and winter mode.    0

DAY cycle, winter mode: The thermostat modulates the heating damper operator when the room temperature is 72°F and below (set point is adjustable). The cooling damper is in the closed position. Between 72 and 75°F (adjustable) room temperature, the thermostat has the heating damper closed and the cooling damper closed or at its minimum open position, whichever is specified. When the room temperature is above 75°F (adjustable), the cooling damper modulates to maintain room temperature and the heating damper remains closed.    350 for 1 VAV unit, 150 for each additional unit on one thermostat

DAY cycle, summer mode: The heating damper is open to a permanently fixed position to provide a minimum constant volume of air flow. The heating side of the thermostat is not in control.    250

When the room temperature is above 75°F (adjustable), the cooling damper operator modulates to maintain room temperature.    0

NIGHT cycle, winter or summer mode: Control for the VAV units is the same as for DAY cycle.    0

MORNING WARMUP cycle, winter or summer mode: Control for the VAV units is the same as for DAY cycle.    0

*Exhaust Fans:* Only those fans with control required by this section should be priced here.

Fans cycled from a reverse-acting room thermostat when the temperature rises above thermostat setting.    250 each

Fans started and stopped from time clock to run on DAY cycle.*    450 each

Fans interlocked to run only when another piece of equipment runs.*    450 each

---

* Cost based on 75-foot wiring run.

**Exhibit 10.1**
**CONTROLS AND INSTRUMENTATION COST ESTIMATING SUMMARY SHEET**

|  | Total cost, $ |
|---|---|
| Compressed air station | _____ |
| Compressed air main(s) | _____ |
| Boiler(s) | _____ |
| Combustion air preheater | _____ |
| Heat exchanger(s) (steam) | _____ |
| Heat exchanger(s) (water) | _____ |
| Scheduled supply fluid | _____ |
| Heating pump(s) | _____ |
| Cooling pump(s) | _____ |
| Hot-water unit heaters and cabinet heaters | _____ |
| Radiation, radiant panels, and convectors | _____ |
| Domestic hot water | _____ |
| Chiller plant | _____ |
| Cooling tower freeze protection | _____ |
| Freeze detection | _____ |
| Smoke detection | _____ |
| HV unit(s) | _____ |
| HV multizone unit(s) | _____ |
| HVAC unit(s) | _____ |
| AC unit(s) | _____ |
| HVAC multizone unit(s) | _____ |
| VAV fan system(s) | _____ |
| VAV fan system(s), cooling only | _____ |
| Dual-duct VAV system(s), heating-side fan | _____ |
| Fan coils | _____ |
| Unit ventilators | _____ |
| VAV units | |
|   Shutoff type | _____ |
|   Shutoff with reheat coils type | _____ |
|   Fan-powered type | _____ |
|   Dual-duct type | _____ |
| Exhaust fans | _____ |
| Miscellaneous_____ | _____ |
| Portion of estimate, this section | _____ |

# 11

# Balancing

The balancing for projects researched was done by independent firms doing only testing and balancing on mechanical projects. Specifications required a professional engineer to be in charge of the work. Typically, the balancing firm had five major responsibilities on the project: (1) reading and recording nameplate data; (2) checking the electrical current to the unit and its starter, and checking the thermal overload heaters for proper size and changing the heaters if necessary; (3) testing, adjusting, and balancing the air, hydronic, and refrigeration systems; (4) testing and adjusting temperature controls for proper operation; and (5) preparing a written report with a copy of the mechanical drawings keyed to the report. In the last few years, testing sound and vibration levels has been included in the balancing firm's work by some engineers. Our specifications have not required this, and the costs do not reflect having done this part of the work.

You must add the balancing contractor's profit to the costs shown in this chapter.

## The Conceptual Estimate

We have compared the cost of balancing to the cost of the heating and cooling portions of the mechanical contracts over the past years. These portions are represented by Chapter 7, Heating and Cooling; Chapter 8, Packaged Heating, Cooling, and HVAC Units; and Chapter 9, Air Distribution Equipment, in this book. Once you have arrived at a cost for these chapters you can use Figure 11.1, which shows the cost of balancing as a percentage of these combined costs. Apply this percentage to these combined costs before assigning a profit to them.

CURVE MAY BE EXTRAPOLATED AT 1.5% FOR LARGER JOBS.

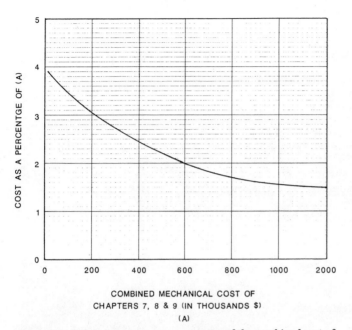

COMBINED MECHANICAL COST OF
CHAPTERS 7, 8 & 9 (IN THOUSANDS $)

(A)

**Figure 11.1** Balancing costs as a percentage of the combined cost of the mechanical work in Chapters 7 to 9 before adding the contractor's profit.

## More Detailed Estimate

After the preliminary drawings stage or the design development stage is completed, the estimator may want to make a more detailed estimate of the balancing costs. We have prepared Table 11.1 to assist in this effort. Table 11.1 shows an outline of the work to be done under each item heading. The work contemplated in the testing, adjusting, and balancing is generally as recommended by the *ASHRAE Handbook*, 1984, Chapter 37, and *National Standards of Associated Air Balance Council*, fourth edition, 1982.

TABLE 11.1    Cost of Balancing Equipment, Ducts, and Piping Systems

| | Hours each | Cost $/each | No. of items | Total cost, $ |
|---|---|---|---|---|
| Boiler:<br>Nameplate data<br>Electrical data<br>Set water temperature operating and<br>  high-limit and safety controls<br>Check gas pressure or fuel oil supply<br>Check relief valve operation<br>Combustion test<br>Record data | 7.5 | 250 | ___ | ___ |
| Pumps, each, heating, cooling, condensing:<br>Nameplate data<br>Electrical data<br>Set flow<br>Record data | 1.5 | 100 | ___ | ___ |
| Antifreeze solution, each heating and<br>cooling:<br>Nameplate data on feeder<br>Electrical data<br>System concentration test<br>Record data | 1.5 | 50 | ___ | ___ |
| Heating mains, each location:<br>Set flows at primary-secondary bridles<br>Record data | .75 | 25 | ___ | ___ |
| Set flows at main balancing valves<br>Record data | .75 | 25 | ___ | ___ |
| Small reheat coils, each:<br>Set flow<br>Check thermostat and valve operation<br>Record data | 1.25 | 41 | ___ | ___ |
| Unit heaters and cabinet heaters, each:<br>Nameplate data<br>Electrical data<br>Measure fan cfm, set belt drive<br>Check thermostat operation<br>Set coil flow<br>Record data | 1.0 | 60 | ___ | ___ |
| Convectors and radiation zone, each:<br>Set water flow<br>Check thermostat and valve operation<br>Record data | 1.0 | 36 | ___ | ___ |
| Electric radiation zone, each:<br>Nameplate and electric data<br>Check thermostat operation<br>Record data | .75 | 25 | ___ | ___ |

TABLE 11.1    Cost of Balancing Equipment, Ducts, and Piping Systems (Continued)

| | Hours each | Cost, $/each | No. of items | Total cost, $ |
|---|---|---|---|---|
| Chiller, each: | 4.0 | 132 | —— | —— |
|   Nameplate data | | | | |
|   Electrical data | | | | |
|   Set water temperatures | | | | |
|   Record flows and pressure drops | | | | |
|   Record data | | | | |
| Air-cooled condenser, each: | 2.00 | 66 | —— | —— |
|   Nameplate data | | | | |
|   Electrical data | | | | |
|   Measure fan cfm | | | | |
|   Check condensing temperatures | | | | |
|   Record data | | | | |
| Cooling towers, each: | 4.5 | 150 | —— | —— |
|   Nameplate data | | | | |
|   Electrical data | | | | |
|   Measure fan cfm | | | | |
|   Check water temperature controls operation | | | | |
|   Check water flow, water nozzles operation, water feed, and water level | | | | |
|   Record data | | | | |
| Condensing water mains, each location: | .75 | 25 | —— | —— |
|   Set flows at main balancing valves | | | | |
|   Record data | | | | |
| Chilled water mains, each location: | | | | |
|   Set flows at primary-secondary bridles | .75 | 25 | —— | —— |
|   Record data | | | | |
|   Set flows at main balancing valves | .75 | 25 | —— | —— |
|   Record data | | | | |
| Fan coils, each: | 1.0 | 66 | —— | —— |
|   Nameplate data | | | | |
|   Electrical data | | | | |
|   Check speed setting and cfm | | | | |
|   Set water flow(s) | | | | |
|   Check thermostat and valve operation | | | | |
|   Record data | | | | |
|   For units with two coils | 2.0 | 72 | —— | —— |
|   Add, for units with automatic outdoor air damper, $16 | | | | |
| Unit ventilators, each: | 2.0 | 83 | —— | —— |
|   Nameplate data | | | | |
|   Electrical data | | | | |
|   Check cfm, set belt drive | | | | |
|   Set water flow(s) | | | | |

TABLE 11.1    Cost of Balancing Equipment, Ducts, and Piping Systems (Continued)

| | Hours each | Cost, $/each | No. of items | Total cost, $ |
|---|---|---|---|---|
| Check thermostat and valve or bypass operation<br>Check outdoor air–return air control operation<br>Record data | | | | |
| Units with two coils | 2.5 | 83 | ____ | ____ |
| HV units, each (coils separate):<br>Nameplate data<br>Electrical data<br>Check cfm, set belt drive<br>Check outdoor air–return air control operation<br>Check thermostat and valve operation<br>Record data | 3.0 | 165 | ____ | ____ |
| For multizones, add 1.0 hr per zone<br>   Typical eight-zone unit | 11.0 | 429 | ____ | ____ |
| AC units, each (coils separate):<br>Balancing is the same as for HV units | 3.0 | 165 | ____ | ____ |
| HVAC units, each (coils separate):<br>Balancing is the same as for HV units | 3.0 | 165 | ____ | ____ |
| For two-deck multizones add 1.0 hr per zone<br>   Typical eight-zone unit | 11.0 | 429 | ____ | ____ |
| For three-deck multizones add 1.3 hr per zone<br>   Typical eight-zone unit | 13.4 | 508 | ____ | ____ |
| Coil banks, each heating and each cooling coil in the unit or coil bank:<br>HV, AC, HVAC unit number<br>Set water flow<br>Record data | 1.1 | 36 | ____ | ____ |
| Furnaces or unit heaters, each:<br>Nameplate data<br>Electrical data<br>Measure fan cfm, set drive or motor tap<br>Check gas pressure or fuel oil supply<br>Check thermostat and high limit operation<br>Record data | 2.0 | 66 | ____ | ____ |
| Condensing units and coils, each:<br>Nameplate data, condenser and coil<br>Electrical data, condenser and coil<br>Measure fan cfm<br>Check condensing temperature<br>Check evaporator temperature<br>Check thermostat-compressor operation<br>Record data | 2.25 | 74 | ____ | ____ |

TABLE 11.1    Cost of Balancing Equipment, Ducts, and Piping Systems (Continued)

|  | Hours each | Cost, $/each | No. of items | Total cost, $ |
|---|---|---|---|---|
| Packaged rooftop cooling unit, each:<br>Nampelate data (each section)<br>Electrical data (each section)<br>Measure evaporator fan cfm, set drive<br>Measure condenser fan(s) cfm<br>Check condensing temperature<br>Check evaporator temperature<br>Check thermostat-compressor operation<br>Manual opposed-blade damper (OBD)<br>Check outdoor air manual OBD setting<br>Record data | 4.0 | 132 | —— | —— |
| For units with outdoor air–return air<br>economizer | 4.5 | 149 | —— | —— |
| Packaged rooftop heating and cooling units,<br>each:<br>All costs of cooling unit plus<br>Nameplate data, heating section<br>Electrical data, heating section<br>Check gas pressure<br>Check heating side of thermostat<br>operation<br>Record data | 5.25 | 190 | —— | —— |
| For units with outdoor air–return air<br>economizer | 6.25 | 190 | —— | —— |
| Evaporative coolers, each:<br>Nameplate data<br>Electrical data<br>Check water supply, water level,<br>overflow,  bleed-off, and drain<br>Check pump operation<br>Check fan cfm, set drive<br>Check operating controls<br>Record data | 2.5 | 107 | —— | —— |
| Makeup air units, each:<br>All costs of evaporative cooler plus<br>Nameplate data, heating section<br>Electrical data, heating section<br>Check gas pressure<br>Check heating side of operating controls<br>Record data | 3.5 | 150 | —— | —— |
| Fans, centrifugal, axial, cabinet<br>vaneaxial, belt-drive propeller, each supply<br>and return fan<br>Nameplate data<br>Electrical data<br>Check cfm, set drive or pitch angle on<br>vaneaxial fans<br>Record data | 2.5 | 83 | —— | —— |

TABLE 11.1  Cost of Balancing Equipment, Ducts, and Piping Systems (Continued)

|  | Hours each | Cost, $/each | No. of items | Total cost, $ |
|---|---|---|---|---|
| For fans with inlet vanes | 3.5 | 115 | ____ | ____ |
| Exhaust fans with belt drive, each:<br>Nameplate data<br>Electrical data<br>Check cfm and backdraft dampers, set<br>    drive<br>Record data | 2.0 | 58 | ____ | ____ |
| Exhaust fans and propeller fans, direct<br>drive, each:<br>Nameplate data<br>Electrical data<br>Check cfm and backdraft dampers<br>Record data | 1.25 | 41 | ____ | ____ |
| Ductwork, each location:<br>Set opposed-blade balancing dampers and<br>    measure cfm in each branch<br>Record data | 0.6 | 41 | ____ | ____ |
| Outdoor air, return air, and exhaust air<br>dampers in built-up fan rooms, each<br>damper:<br>Measure cfm<br>Check operation<br>Record data | 1.0 | 33 | ____ | ____ |
| Variable air volume or constant volume<br>control units, each:<br>Nameplate data<br>Measure and set cfm<br>Check thermostat and damper operation<br>Record data | 1.25 | 41 | ____ | ____ |
| For double-duct units | 1.50 | 50 | ____ | ____ |
| Fan-powered units, each:<br>Nameplate data<br>Electrical data<br>Measure and set primary cfm<br>Measure and set fan cfm<br>Check thermostat and damper operation<br>Record data | 1.25 | 58 | ____ | ____ |
| Add for water reheat coils, each | 1.0 | 16 | ____ | ____ |
| Add for electric reheat coils, each | 0.5 | 8 | ____ | ____ |
| Registers, each:<br>Measure and set cfm<br>Record data | 0.5 | 25 | ____ | ____ |

TABLE 11.1    Cost of Balancing Equipment, Ducts, and Piping Systems (Continued)

|  | Hours each | Cost, $/each | No. of items | Total cost, $ |
|---|---|---|---|---|
| Diffusers, square, rectangular, or round, each:<br>Measure and set cfm<br>Record data | 0.5 | 25 | ——— | ——— |
| Linear diffusers, 8-ft length<br>Measure and set cfm<br>Record data | 1.0 | 33 | ——— | ——— |
| Slot diffusers and troffer outlets, 4-ft length, each:<br>Measure and set cfm<br>Record data | 0.5 | 25 | ——— | ——— |
| Exhaust register:<br>Measure and set cfm<br>Record data | .58 | 20 | ——— | ——— |
| Bound reports (three copies each):<br>Data and Balancing report with drawings coded to report: |  |  |  |  |
| Smallest jobs | 7 to 8 | 250 | ——— | ——— |
| Small jobs | 16 to 20 | 600 | ——— | ——— |
| Average jobs | 30 | 1,000 | ——— | ——— |
| Large jobs | 44 | 1,500 | ——— | ——— |

# Chapter

# 12

# Insulation

The insulation on projects researched was accomplished by independent insulating firms doing the work on a subcontract basis for the mechanical contractor.

## The Conceptual Estimate

We have compared the cost of insulation to the cost of the plumbing and heating-cooling portions of the mechanical contracts over the past years. These portions are represented by Chapter 6, Plumbing, and Chapter 7, Heating and Cooling. Once you have arrived at a cost of these two chapters you can use Figure 12.1, which shows the cost of the insulation contract (for insulating the piping and equipment) as a percentage of the combined costs of these two chapters. Apply the percentage to these combined costs before assigning the contractor's profit to them. Curves 1 and 2 show the lower and upper limits of projects researched. No key could be found to explain why the insulation contracts were smaller on some jobs than on others. On projects where packaged heating-cooling units are used and there is no piping or equipment insulation under the heating-cooling section, the cost from Figure 12.1 can be applied as a percentage of the *Plumbing* cost only.

On the projects researched, *all or most of the ductwork was rectangular with interior insulation liner, furnished and installed by the sheet metal contractor. If the project has ductwork insulated on the outside, the estimator should add the cost of this insulation to the cost of the unlined ductwork from Chapter 9 and to the cost from Figure 12.1.* Use

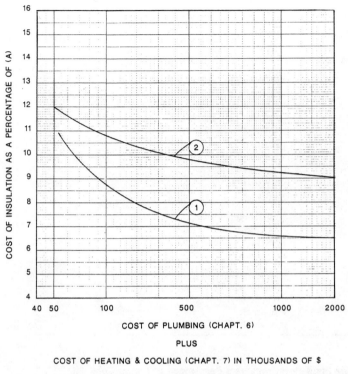

COST OF PLUMBING (CHAPT. 6)

PLUS

COST OF HEATING & COOLING (CHAPT. 7) IN THOUSANDS OF $

(A)

**Figure 12.1**  Insulation costs as a percentage of the combined cost of the mechanical work in Chapters 6 and 7 before adding the contractor's profit. (1) Lower limit; (2) upper limit.

Table 12.1 for these costs. Also shown in Table 12.1 are costs in dollars per square foot of surface for insulating various pieces of hot and cold equipment for use when you are doing a material and labor cost estimate.

## More Detailed Estimate

After the preliminary drawings stage or the design development stage is completed, the estimator may want to make a more detailed estimate of the insulation costs. For this we have prepared Table 12.2, which shows the costs of pipe insulation installed as a percentage of piping systems. The total piping cost, including pipe, valves, and fittings, should be used when applying these percentages (see Chapter 7). Insulation is the glass fiber type with a vapor barrier jacket on cold

TABLE 12.1  Equipment and Duct Insulation Costs

| Item | Insulation | Cost, $/ft$^2$ surface | Cost for supply ducts, $/cfm |
|---|---|---|---|
| Boilers, hot water; storage tanks; heat exchangers; breeching | $1\frac{1}{2}$-in calcium silicate, $\frac{1}{2}$-in cement, 8-oz canvas, fire-resistive mastic | 14.00 | |
| Pump casings and cold equipment | $\frac{3}{4}$-lb, 2-in-thick glass fiber blanket, glass fiber mesh, vapor sealer, 8-oz canvas, fire-resistive mastic | 8.75 | |
| Ductwork | $\frac{3}{4}$-lb glass fiber blanket with foil-kraft vapor barrier facing: | | |
| | 1 in thick | 1.00 | .39 |
| | 2 in thick | 1.50 | .48 |
| | $\frac{3}{4}$-lb blanket with no facing: | | |
| | 1 in thick | .80 | .32 |
| | 2 in thick | 1.30 | .42 |
| | 3-lb rigid glass fiber board with foil-kraft vapor barrier facing: | | |
| | 1 in thick | 4.90 | |
| | 2 in thick | 5.90 | |

TABLE 12.2  Pipe Insulation Costs as a Percentage of Piping Cost

| Pipe size, in | Hot* and cold copper water pipe Insulation thickness, in | % | Chilled water pipe Insulation thickness, in | % | Hot water heating (201–250°) pipe Insulation thickness, in | % | Galvanized steel drainage pipe Insulation thickness, in | % |
|---|---|---|---|---|---|---|---|---|
| $\frac{1}{2}$ | $\frac{1}{2}$ | 30 | $\frac{1}{2}$ | 30 | 1 | 33 | — | — |
| $\frac{3}{4}$ | $\frac{1}{2}$ | 30 | $\frac{1}{2}$ | 28 | 1 | 30 | — | — |
| 1 | $\frac{1}{2}$ | 27 | $\frac{1}{2}$ | 26 | 1 | 27 | — | — |
| $1\frac{1}{4}$ | $\frac{1}{2}$ | 26 | 1 | 27 | $1\frac{1}{2}$ | 36 | — | — |
| $1\frac{1}{2}$ | $\frac{1}{2}$ | 26 | 1 | 27 | $1\frac{1}{2}$ | 36 | — | — |
| 2 | $\frac{1}{2}$ | 21 | 1 | 23 | $1\frac{1}{2}$ | 30 | $\frac{1}{2}$ | 20 |
| $2\frac{1}{2}$ | $\frac{1}{2}$ | 18 | 1 | 20 | $1\frac{1}{2}$ | 27 | $\frac{1}{2}$ | 20 |
| 3 | $\frac{1}{2}$ | 17 | 1 | 20 | $1\frac{1}{2}$ | 25 | $\frac{1}{2}$ | 19 |
| 4 | $\frac{1}{2}$ | 13 | 1 | 18 | $1\frac{1}{2}$ | 22 | $\frac{1}{2}$ | 18 |
| 5 | — | — | 1 | 15 | 2 | 19 | $\frac{1}{2}$ | 17 |
| 6 | — | — | 1 | 15 | 2 | 19 | $\frac{1}{2}$ | 16 |
| 8 | — | — | 1 | 15 | 2 | 17 | $\frac{1}{2}$ | 15 |

* Numbers shown here are for vapor barrier jacketed insulation for cold service and with all valves covered. For insulation jacketed for hot water pipe, the cost is 1 to 2 percentage points less, if the valves are covered. If the valves are not covered, the cost may be 5 percentage points less.

**Exhibit 12.1**
**INSULATION COST ESTIMATE SUMMARY SHEET**

|  | Total cost,$ |
|---|---|
| Cold water piping | _____ |
| Hot water piping | _____ |
| Chilled water piping | _____ |
| Heating water piping | _____ |
| Roof drainage piping | _____ |
| Boiler(s) | _____ |
| Hot water storage tanks | _____ |
| Heat exchangers | _____ |
| Breeching | _____ |
| Pump casings | _____ |
| Cold equipment | _____ |
| Ductwork | _____ |
| Miscellaneous_____ | _____ |
| Portion of estimate, this section | _____ |

pipe and all-purpose jacket on hot pipe. Table 12.1 also shows the costs of insulating equipment and ducts.

Last, we have an insulation cost summary sheet, Exhibit 12.1, to use with the more detailed estimate.

Add the insulation contractor's profit to these figures to get the total subcontract price to the mechanical contractor.

# 13

# Fire Protection

The fire protection systems discussed in this chapter are installed and sometimes designed by a fire protection contractor. Included are wet sprinkler and piping systems, fire entry stations, zone control valve assemblies, fire pump assemblies, dry pipe systems, standpipe systems, Halon fire protection, and kitchen rangehood chemical extinguisher systems.

All piping 2 in and below was priced as screwed pipe, valves, and fittings; pipe $2\frac{1}{2}$ in and above was priced using grooved piping type couplings and fittings (Table 13.1).

## Wet Sprinkler and Piping Systems

The curves in Figures 13.1 and 13.2 show the cost in dollars per square foot based on an estimated or actual number of square feet per sprinkler head. The curve in Figure 13.1 should be used for ordinary hazard systems where the number of heads per square foot can be expected to range between 103 and 111. The curve in Figure 13.2 should be used for light hazard systems where the number of heads per square foot can be expected to range between 130 and 180. The number of sprinkler heads per square foot is dependent on the floor plan. A building with numerous small offices will have a much greater number of sprinkler heads per square foot than a building with large areas of open space. The costs of the fire entry station or stations, any zone control valves, dry pipe, standpipes, Halon, or kitchen rangehoods, or other type of fire protection system should by added to the costs shown in Figures 13.1 and 13.2.

All piping used to price these systems was schedule 40 black steel. Prices for piping and welded or screwed fittings can be found in Chapter 7, Heating and Cooling. Each sprinkler head was priced as an

TABLE 13.1    Grooved Pipe Fittings Cost

| Steel pipe size, in | Cost, $/each installed | | |
|---|---|---|---|
| | 90° elbows | Tees | Unions |
| $2\frac{1}{2}$ | 59 | 87 | 31 |
| 3 | 69 | 105 | 36 |
| 4 | 97 | 148 | 51 |
| 5 | 151 | 236 | 70 |
| 6 | 190 | 292 | 82 |
| 8 | 312 | 489 | 132 |

exposed chrome-plated head with a one-piece chrome escutcheon plate (Table 13.2). The systems were assumed to be installed in average 8- to 10-ft ceiling heights, with adequate ceiling space available to run piping. The 1986 bidding climate has forced the price of fire protection systems down in some parts of the country. The effects of this bidding climate vary with location and have not been addressed in the prices given in this chapter. Estimators may want to adjust the cost estimate to include the effects of the bidding climate in their area.

## Fire Entry Station

Figure 13.3 details a fire entry station. For the purposes of this chapter, prices were compiled for both 4- and 6-in fire entry stations. Fifteen feet of piping was allowed for each pipe to a drain. Piping to the

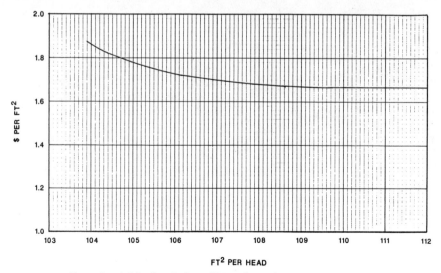

Figure 13.1    Cost of sprinkler heads for ordinary hazard systems.

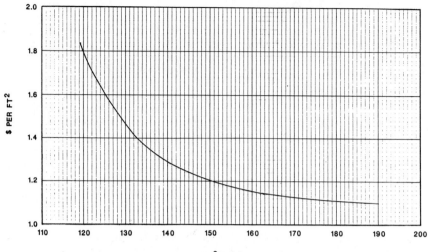

FT² PER HEAD

**Figure 13.2** Cost of sprinkler heads for light hazard systems.

**TABLE 13.2 Sprinkler Head Cost**

| Type | Cost, $/each installed |
|---|---|
| Chrome-plated pendent | 13.40* |
| Rough brass pendent or chrome-plated sidewall | 12.60* |
| Chrome escutcheon plate: | |
|   Two-piece adjustable size | 2.10 |
|   One piece | .40 |

* Price includes cutting 1-in supply pipe and threading it to match ceiling or mounting height.

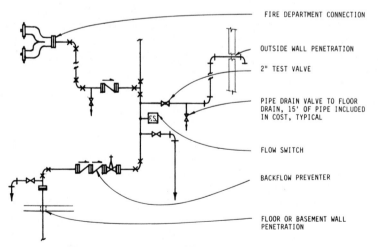

FIRE DEPARTMENT CONNECTION

OUTSIDE WALL PENETRATION

2" TEST VALVE

PIPE DRAIN VALVE TO FLOOR DRAIN, 15' OF PIPE INCLUDED IN COST, TYPICAL

FLOW SWITCH

BACKFLOW PREVENTER

FLOOR OR BASEMENT WALL PENETRATION

**Figure 13.3** Fire entry station assembly.

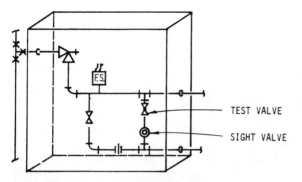

TEST VALVE

SIGHT VALVE

**Figure 13.4** Zone control valve assembly.

fire department connection and to the outside wall penetration should be added to the estimate. Also included, but not shown in Figure 13.3, is the price of an alarm bell. The cost of a 4-in fire entry station is $4,505. The cost of a 6-in fire entry station is $7,255.

### Zone Control Valve Assembly

Figure 13.4 details a zone control valve assembly in a cabinet. The zone is priced with $2\frac{1}{2}$-in pipe and includes the connection to the fire main and supply and drain piping to outside the cabinet. The alarm system should be connected to the flow switch; 100 ft of low-voltage control wiring in conduit is included in the estimate. The price of a zone control valve assembly is $1,570. This includes $200 for a steel cabinet with a steel door.

### Standpipes and Hose Cabinets

Figure 13.5 details a standpipe hose valve in a cabinet. Two different sizes of hose valves are generally available: $2\frac{1}{2}$-in valves are for fire department use and $1\frac{1}{2}$-in valves are for occupant use. There are several options and combinations of equipment available. All combinations are priced with steel cabinets and full glass doors.

| | |
|---|---|
| $1\frac{1}{2}$-in hose valve, hose and cabinet | $ 750 |
| $2\frac{1}{2}$-in hose valve and cabinet (hose provided by fire department) | $ 545 |
| Add—For large cabinet and 10-lb dry chemical fire extinguisher | $ 150 |

The price for the hose cabinet includes a tee off of the standpipe riser. The standpipe costs should be added to the system costs. Costs for vertical and horizontal piping are listed in Chapter 7, Heating and Cooling, and should be used to estimate the cost of the standpipe mains.

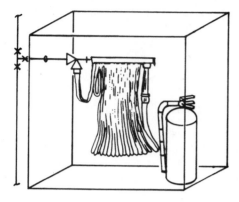

**Figure 13.5** Fire hose cabinet assembly.

## Fire Pump Assemblies

Figure 13.6 details an electric-driven fire pump assembly. Systems were priced for 75, 150, 200, and 275 ft of head, which correspond to 5-, 10-, 15-, and 20-story buildings. Jockey pumps were assumed to have a capacity of 1%, that of the fire pump. If emergency electrical service is required, that cost should be added to the costs given here. The prices for the fire pump systems include across-the-line starters and controllers for all pumps. Figure 13.7 shows cost per gallon per minute vs. the fire pump gallon per minute rate for 75-ft and 275-ft pumping heads. The estimator should know that while pump costs for each pumping head curve increase 50% from 250 to 1,500 gpm, the system

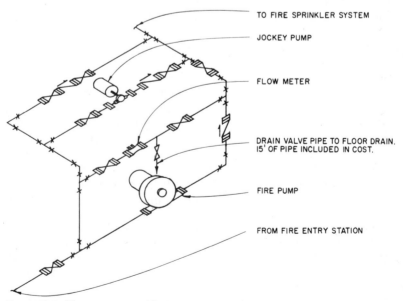

TO FIRE SPRINKLER SYSTEM

JOCKEY PUMP

FLOW METER

DRAIN VALVE PIPE TO FLOOR DRAIN.
15' OF PIPE INCLUDED IN COST.

FIRE PUMP

FROM FIRE ENTRY STATION

**Figure 13.6** Fire pump assembly.

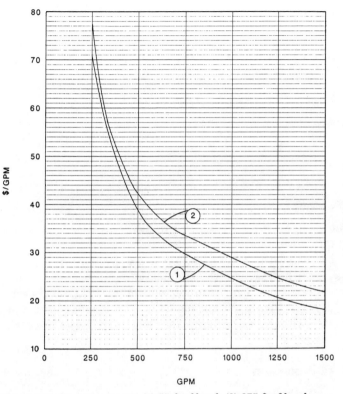

**Figure 13.7**  Fire pump costs. (1) 75 ft of head; (2) 275 ft of head.

cost is heavily dependent on pipe size. The following pipe sizes were used at their corresponding flow rates: 250 gpm—4 in, 500 gpm—5 in, 750 gpm—6 in, 1,000 and 1,500 gpm—8 in.

## Dry Pipe Sprinkler Valve Assembly

Dry pipe sprinkler valve assembly principally includes a dry pipe sprinkler valve and an air compressor. The air compressor should be sized as per National Fire Protection Association (NFPA), Chapter 13. The dry pipe sprinkler valve has been priced in $2\frac{1}{2}$-, 3-, and 4-in sizes, which includes the price of the dry pipe valve, a gate valve on the wet side, and a trim package including pressure gauges, fill cups, etc. Since the length of pipe required from the air compressor to the dry pipe valve varies with each application, that cost should be added to the cost of the assembly listed below. Usually $\frac{1}{2}$-in copper pipe is used and 25 ft of that is included in the dry sprinkler valve estimate. To determine the cost of the dry pipe sprinkler assembly, add the cost of the appropriate air compressor to the dry pipe sprinkler valve cost shown below:

| Air compressor, hp | Cost, $ |
|---|---|
| $\frac{1}{3}$ | 1,006 |
| $\frac{1}{2}$ | 1,301 |
| $\frac{3}{4}$ | 1,441 |
| 1 | 1,548 |
| 5 | 2,212 |

| Dry sprinkler valve, in | Cost, $ |
|---|---|
| 2 | 896 |
| $2\frac{1}{2}$ | 1,000 |
| 3 | 1,100 |
| 4 | 1,282 |

## Halon Fire Protection

Halon fire protection systems are difficult to estimate due to the many factors which influence their design. These factors seem to level off as the jobs get larger. Costs were developed for systems protecting 10-ft-high spaces including underfloor detection and release. Figure 13.8 shows costs per cubic foot of protected volume. Halon systems include storage tanks, piping, heads, detection and alarm equipment, abort and release switches, an annunciator panel, and one release test using Halon as the test gas.

## Rangehood Fire Protection

Costs have been developed for rangehood fire protection systems using an R102 or R101 wet or dry chemical extinguisher. The system consists

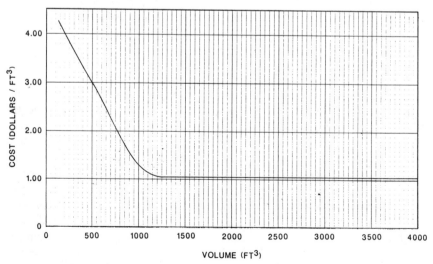

**Figure 13.8** Halon systems costs.

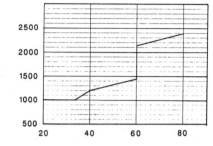

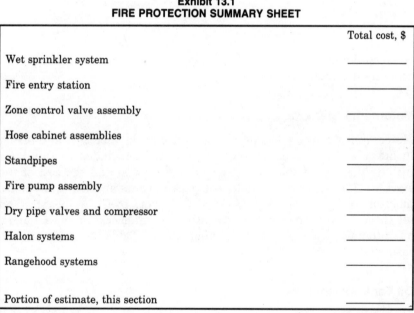

Figure 13.9 Rangehood fire protection costs.

RANGE HOOD SIZE (FT$^2$)

Exhibit 13.1
FIRE PROTECTION SUMMARY SHEET

|  | Total cost, $ |
|---|---|
| Wet sprinkler system | _____ |
| Fire entry station | _____ |
| Zone control valve assembly | _____ |
| Hose cabinet assemblies | _____ |
| Standpipes | _____ |
| Fire pump assembly | _____ |
| Dry pipe valves and compressor | _____ |
| Halon systems | _____ |
| Rangehood systems | _____ |
| Portion of estimate, this section | _____ |

of the chemical, a tank, piping, sensors, and distribution nozzles. The system has been priced for 4-ft-deep rangehoods with standard cooking equipment underneath. Any equipment which requires special protection will require that additional costs are added to those shown in Figure 13.9. Sharp adjustments in the curve occur where major system components, mainly tanks of extinguisher, are added to cover the additional area. The exact location of those sharp adjustments in the curve is dependent on the condition of each hood, including the types of cooking equipment being used.

**Summary**

Exhibit 13.1 is the fire protection summary sheet.

# 14

# Special Systems

Some equipment and their assemblies which are not included in other chapters are included here.

## Motors and Starters

Motors and starters are generally included in the mechanical contract in many areas of the country. All the prices included in this book include the cost of the motor and its installation and the cost of the starter. Starter costs include the starter and enclosure, a phase monitor relay on starters 5 hp and larger, overload heaters (on all three legs of three-phase starters), an interlock relay, a hands-off-automatic switch, and the pilot light. Table 14.1 shows the costs used.

## Oil Tank Assemblies

Shown in Figure 14.1 are the dollar costs per tank gallon for buried glass fiber fuel oil tanks. Concrete pads or supports are not included. All of the following items are included and are suitable for no. 2 fuel oil storage. Costs are for the installation in commercial projects.

Excavation

Tank with UL label and tiedowns with tank manhole cover, ladder, and fittings

Surround tank with 6 in sand

Backfill, compacted in 12-in layers

**TABLE 14.1    Motor and Starter Costs**

| Motor size, hp | Motor cost, $ | | | NEMA size/maximum hp across the line | | Starter cost, $ | |
| | Open drip-proof | Explosion-proof | Instal-lation* | Size | 208 V | 208 V | 440 V |
| --- | --- | --- | --- | --- | --- | --- | --- |
| ½ | 105 | 185 | 135 | 00 | 1½ | 135 | — |
| ¾ | 115 | 200 | 135 | 00 | 1½ | 135 | — |
| 1 | 120 | 230 | 225 | 00 | 1½ | 135 | — |
| 1½ | 135 | 245 | 225 | 00 | 1½ | 135 | — |
| 2 | 145 | 255 | 225 | 0 | 3 | 155 | — |
| 3 | 165 | 295 | 225 | 0 | 3 | 155 | — |
| 5 | 185 | 360 | 225 | 1 | 7½ | 280 | — |
| 7½ | 255 | 490 | 225 | 1 | 7½ | 280 | — |
| 10 | 300 | 575 | 225 | 2 | 10 | 390 | — |
| 15 | 395 | 820 | 225 | 3 | 25 | 545 | — |
| 20 | 490 | 980 | 405 | 3 | 25 | 545 | — |
| 25 | 610 | 1,260 | 405 | 3 | 25 | 545 | — |
| 30 | 700 | 1,440 | 630 | | 50 | 1,700 | 1,220 |
| 40 | 875 | 1,935 | 630 | | 50 | 1,700 | 1,220 |
| 50 | 1,055 | 2,220 | 630 | | 50 | 1,700 | 1,220 |
| 60 | 1,345 | — | 630 | | 75 | 3,370 | 1,465 |
| 75 | 1,630 | — | 810 | | 75 | 3,370 | 1,465 |
| 100 | 2,150 | — | 810 | | 150 | 5,930 | 2,625 |

* Labor cost when installed in field, not at factory.

Hauling excess dirt 5 mi

Rental of backhoe and dump truck

3-in fill and sounding lines and locking caps, and 9-in, round, cast iron manholes with gasketed covers placed in 18-in-square concrete pad

1-in fuel oil supply (FOS) and fuel oil return (FOR) lines to the edge of the excavation with four ells each

Angle check valve or antisiphon valve

Suction line strainer

$1\frac{1}{2}$-in vent line to the edge of the excavation with four ells

Liquid level gauge with remote readout 100 ft away

All piping coated and wrapped

For buried piping laid on a 6-in bed of sand and covered by 6 in of

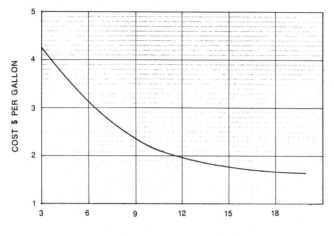

TANK SIZE IN 1000 GALLONS

**Figure 14.1** Cost of buried fuel oil tank assemblies.

sand in a trench 4 ft wide and 3 ft deep, compacted by hand in 12-in layers, the cost is $7.20 per foot plus piping costs. For 1-in FOS and FOR and a 1½-in vent laid in the trench, the cost is $18.95 per foot, or a total cost of $26.15 per foot for oil piping and vent from the tank to the boiler room. These sizes are suitable for no. 2 fuel oil.

## Solar Heating Assemblies

Costs of projects installed by the Solaron Company, Englewood, Colorado, as well as projects designed by our office from 1975 through 1981, were assembled. Table 14.2 shows the average installed costs of solar heating assemblies for new residential and for both new and retrofitted

**TABLE 14.2  Solar Heating Assemblies**

| System type | Cost of gross collector area, $/ft² |
|---|---|
| Residential: | |
| Domestic hot water | 51 |
| Space heating with liquid collectors and heating media | 42 |
| Space heating with air collectors and heating media | 40 |
| Commercial: | |
| Domestic or process hot water | 51 |
| Space heating with liquid collectors and air heating | 60 |
| Space heating with air collectors and heating media | 47 |

commercial installations. All projects used factory-made flat plate collectors.

Costs are for the solar portion complete, including contractor's overhead and profit.

Controls, instrumentation, size, and type of storage as well as many other factors can vary these costs greatly.

### Indirect-Direct Evaporative Cooling Assemblies

Three systems were priced, and their costs are shown on Figure 14.2. The systems were

1. Z-Duct IDC Units, Des Champs Laboratories, East Hanover, New Jersey, for connection to fan and cooling system. The Z-Duct unit is complete with sump, pump, float valve assembly, scavenger air fan(s), motor(s) and starter(s), indirect cooler heat exchanger, and direct evaporative cooler with 12-in-deep media. Makeup water; bleed, overflow, and drain piping for manual draining; and labor to install the entire assembly were included.

Assumed conditions were 100% outdoor air cooling, 95° dry bulb, 60° wet bulb. Indirect efficiency was 80%; direct evaporative cooling efficiency was 89%.

2. Sun Class 12 and 10 Air Washers, Sun Manufacturing, Incorporated, El Paso, Texas. Class 10 units were the direct evaporative cooling component for connection to the fan and cooling system, and included the air washer, sump, pump, motor and starter, and float

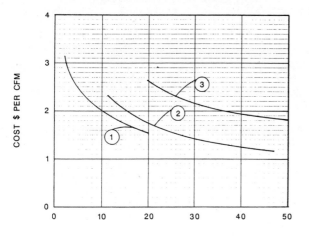

CFM IN 1000'S

**Figure 14.2** Indirect-direct evaporative cooling assemblies. (1) Z-Duct IDC units; (2) Sun Class 12 and 10 air washers; (3) Howden heat pipe units.

valve. Makeup water; bleed, overflow, and drain piping for manual draining, and labor to install the entire assembly were included.

Class 12 units were the indirect component (remotely located 25 ft distant) with 60 ft of piping, balancing valve air vent and drain valve included. Also included were the air washer, sump, recirculating pump to coil, 40 ft of head, motor and starter, float valve, and V-belt-driven propeller fan with backdraft damper. A six-row coil attached to the direct cooler intake was included to complete the indirect portion of the cooling. Makeup water, bleed, overflow, and drain piping for manual draining and labor to install the entire assembly was included.

Assumed conditions were the same as for the other types of coolers. Indirect efficiency was 75%, and the direct evaporative cooling efficiency was 89%.

3. Howden Heat Pipe Units, Howden Heat Pipe Division, Bloomfield, Connecticut, for connection to fan and cooling system. The heat pipe unit is complete with six-row deep heat exchangers, indirect heat pipe cooler, direct evaporative cooler with 12-in-deep media, common sump, pumps, motor(s) and starters, float valve, and eliminators. Makeup water; bleed, overflow, and drain piping for manual draining, and labor to install the entire assembly were included.

# 15

# Cost Estimating Summary

Now that we have arrived at the cost of each portion of the mechanical work, we can summarize these for a total mechanical contract cost. But before we can do this we must modify the costs for our area index, apply an inflation factor to the midterm of construction, evaluate the completeness of our design information to determine what design contingency (if any) to apply, and assign a contractor's profit to each portion. Once this is done, we determine what markup the mechanical contractor would probably apply to each subcontract. By using a multiplier for this number we arrive at the cost to the general contractor. The general contractor's markup for administration typically has been small. Unless you have a better insight, apply 3% for the smallest job and 1% for the largest.

### Multipliers

The area index multiplier can be found in Table 2.1. The inflation factor to use may be found for Colorado from Figure 2.1. If you have similar information for your area, you may want to use that information to arrive at an inflation factor. We have normally applied this inflation factor from the date of the estimate to the anticipated midterm of construction.

Figure 15.1 shows design contingency and profit as job cost multipliers. Design contingency should be based on the total mechanical cost and should only be used when the designer is not sure that all of the design requirements are known at the start of the project. At the time the designer believes all the requirements are known, this contingency factor should be reduced to zero. Profit is applied to *each portion of the work performed by one contractor*. Since each of these

**Exhibit 15.1**
**COST ESTIMATING SUMMARY SHEET**

| Portion of estimate | Cost from summary sheets | Multipliers | | | | | Cost to mechanical contractor | Mechanical contractor's markup multiplier | Cost estimate to general contractor |
|---|---|---|---|---|---|---|---|---|---|
| | | Area index | Inflation factor | Design contingency | Profit, each contractor | Total multiplier | | | |
| Site work (Chapter 5) | | | | | | | | | |
| Plumbing (Chapter 6) | | | | | | | | | |
| Heating and cooling (Chapter 7) | | | | | | | | | |
| Packaged equipment (Chapter 8) | | | | | | | | | |
| Air distribution (Chapter 9) | | | | | | | | | |
| Controls (Chapter 10) | | | | | | | | | |

Balancing
(Chapter 11)

Insulation
(Chapter 12)

Fire protection
(Chapter 13)

Special systems
(Chapter 14)

Other items

Bonds

Total cost estimate of the mechanical systems

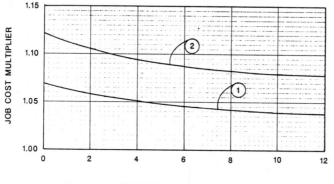

JOB COST IN 100,000 DOLLARS

**Figure 15.1** Design contingency and profit as a percentage of job cost. (1) Design contingency; (2) profit.

subcontracts or portions is a different size, the profit multiplier will be different for each one.

The mechanical contractor's markup should not be applied to the portions which that contractor is performing in house, to which you have already applied a profit. The mechanical contractor will not do all of the work under the contract in house. Some will have plumbers and pipe fitters and do plumbing and heating in house. Others will have their own sheet metal shop and subcontract the insulation, temperature control, etc. The percentage of markup on these subcontracts is difficult to estimate. The author believes that market conditions have as much to do with this as anything. In lean times when a contractor needs work he or she may not mark up subcontractor prices at all. For a guideline, in normal times we use 5 to 8%.

# Bibliography

*ASHRAE Handbook, 1984 Systems*, American Society of Heating, Refrigerating and Air Conditioning Engineers, Inc., Atlanta, Georgia.

*National Standards, Associated Air Balance Council*, fourth edition, Associated Air Balance Council, Washington, D.C., 1982.

Percival E. Pereira, chief editor, *1986 Dodge Manual for Building Construction Pricing and Scheduling Cost Information Systems*, McGraw-Hill Information Systems Company, Princeton, New Jersey.

A. M. Khashab, PE, *Heating, Ventilating, and Air Conditioning Systems Estimating Manual*, second edition. McGraw-Hill Book Company, New York, New York, 1985.

*General Construction Estimating Standards*, volume 3, *Mechanical and Electrical*, Richardson Engineering Service, Inc., San Marcos, California, 1986.

*Labor Estimating Manual*, Mechanical Contractors' Association of America, Inc., Washington, D.C., 1983.

Gordon P. McKinnon, editor in chief, *Fire Protection Handbook*, fifteenth edition, National Fire Protection Association, Quincy, Massachusetts, 1981.

Mellville J. Mossman, editor in chief, *Means Mechanical Cost Data, 1986*, ninth annual edition, Robert Snow Means Company, Inc., Kingston, Massachusetts, 1985.

John Gladstone, *Air Conditioning Testing/Adjusted/Balancing, A Field Practice Manual*, second edition, Van Nostrand Reinhold Company, Inc., New York, New York, 1981.

*Weather Data Handbook*, Ecodyne Cooling Products Division, McGraw-Hill Book Company, New York, New York, 1980.

*ASHRAE Handbook, 1985 Fundamentals*, American Society of Heating, Refrigerating and Air Conditioning Engineers, Inc., Atlanta, Georgia.

*HVAC Duct System Design*, second edition, Sheet Metal and Air Conditioning Contractors National Association, Inc., Vienna, Virginia, 1981.

Roger W. Haines, *Control Systems for Heating, Ventilating and Air Conditioning*, second edition, Van Nostrand Reinhold Company, New York, New York, 1971.

*The Bell & Gossett Booster Pump, Bulletin*, A-107A, ITT Corporation, 1979.

*Fan Application Manual Part 1*, A.M.C.A. Publication 201, Fans and Systems, Air Moving and Conditioning Association, Arlington Heights, Illinois, 1975.

Jerry Pope and Jerry Morgensen, "Colorado Building Construction Cost Increase Averages," Helsel Phelps Construction Company, Greeley, Colorado. Update January 1, 1986.

Project Files, McFall-Konkel & Kimball, Consulting Engineers, Inc., Denver, Colorado, and Cheyenne, Wyoming, 1975–1985.

# Table of Equivalents

| One | Is equal to | One | Is equal to |
|---|---|---|---|
| Foot (ft) | 12 in | Gallon per minute | $6.309 \times 10^{-5}$ cm³/s |
| Inch (in) | 2.54 centimeter (cm) | (gpm, gal/min) | ($6.309 \times 10^{-2}$ L/s) |
| Foot (ft) | 0.3048 meter (m) | Inch of water | 0.03609 psi |
| Cubic foot (ft³) | 0.02832 m³ | (at 60°F) | |
| Cubic yard (yd³) | 0.7646 m³ | Inch of mercury | 0.4912 psi |
| U.S. gallon (gal) | 0.1337 ft³ | (at 32°F) | |
| Liter (L) | 0.03531 ft³ | U.S. horsepower- | 2544.65 Btu |
| Cubic foot | $4.72 \times 10^{-4}$ m³/s | hour (hp · hr) | |
| per minute (ft³/ | ($4.72 \times 10^{-1}$ L/s) | Btu (mean) | 252.0 calories (cal) |
| min) | | Boiler horsepower | 33,475 Btu/hr |
| Ounce (oz) | 0.02835 kilogram (kg) | (hp) | |
| Pound (lb) | 0.4536 kg | U.S. horsepower | 76.04 kg · m/s |
| Foot per minute | 0.00508 m/s | (hp) | |
| (ft/min) | | U.S. horsepower | 747.70 Watts (W) |
| Mile per hour | 1.609 km/hr | (hp) | |
| (mi/hr, mph) | | Boiler horsepower | 9,809.50 W |
| Foot of water (at | 0.43310 pounds per | (hp) | |
| 39.2°F) | square in (psi, lb/in²) | Ton of | 3,516.8 W |
| Atmosphere | 14.70 psi | refrigeration | |
| Atmosphere | 1.0333 kg/cm² | Ton of | 12,000 Btu/hr |
| Kilowatthour | 3412.66 Btu | refrigeration | |
| Square inch (in²) | $6.541 \times 10^{-4}$ m² | Metric horsepower | 75 kg · m/s |
| Square foot (ft²) | 0.0929 m² | Metric horsepower | 735.50W |
| Square yard (yd²) | 0.8361 m² | Btu per hour | 0.2929W |
| U.S. gallon (gal) | 3.785L | (Btu/hr) | |
| Liter (L) | 0.001 m³ | Calories per second | 4.184W |
| | | (cal/s) | |

To convert from Fahrenheit (°F) to Celsius (°C): $°C = \frac{5}{9}(°F - 32)$.

# B

# Mechanical legend

—◁ᵈ▷— FLOW INDICATOR

—▷◁— GATE VALVE

—▷◀— GLOBE VALVE

—ᴺ— CHECK VALVE

—▷ᴺ— FLOW-CONTROL VALVE

—◁◻— PLUG OR BALANCING VALVE

DRAIN VALVE WITH HOSE END

WASTE

— s— SANITARY SEWER

—w— WATER MAIN

—sᴛ— STORM DRAIN

—ꜰ— FIRE LINE

—ɢ— GAS

STRAINER WITH BLOW-OFF VALVE

SAFETY RELIEF VALVE

—◁ᵀ— TEMPERATURE CONTROL VALVE

—◁ᴾ— PRESSURE-REDUCING VALVE

—ꜱ— AIR VENT

—↑— PRESSURE-TEMP. TAP

—Ⓝ— PRESSURE GAUGE WITH PIG TAIL & COCK

⊏=⊐—ꞁ THERMOMETER

Ⓣ THERMOSTAT

—ꜱ— VACUUM BREAKER

—⊗— PIPE ANCHOR

—▨▨▨— FLEXIBLE CONNECTOR

—▮◗◖▮— PUMP & EQUIPMENT CONNECTOR

—ᴠ— PLUMBING VENT

—꜀꜀— PIPE UNION

ꜰ WELDED PIPE FITTINGS

ꜰ SCREWED PIPE FITTINGS

ꜰ GROOVED PIPE FITTINGS

# Index

Air conditioning (AC) systems (*see* Heating, ventilating, and air conditioning systems)
Air conditioning (AC) unit assemblies, 41
Air conditioning (AC) unit controls, 179–180
(*See also* Heating, ventilating, and air conditioning unit controls)
Air conditioning (AC) units, 115–116, 118–133
coil costs for, 124
coil piping assemblies in, 120–133
without coils, 118–120
costs for, 119, 124
multizone and double-duct, 120
(*See also* Cooler assemblies; Coolers; Cooling; *entries beginning with the term*: Cooling)
Air-cooled chiller/condensers, 174
Air-cooled chiller plant controls, 174
Air-cooled chillers, 96–98, 107
Air-cooled condenser assemblies, 99–100
Air-cooled condensers, 174
Air distribution equipment, 46, 147–166
(*See also* Air moving equipment)
Air ducts, 44, 157–158
Air fitting, 71
Air heater assemblies, 36, 72
Air lines, main, 169
Air louvers, 71, 158, 159
Air moving equipment, 42–47, 164
(*See also* Air distribution equipment; Fans)
Air preheater controls, 170
Air quantities, 20
Air returns, 44, 45, 164, 181–185
Air separation systems, 71, 98
Air stations, compressed, 169
Air supply, 33–34

Air temperature, 82
(*See also* Temperature controls)
Air units, make-up, 146
Altitude in cost estimating, 3
Antifreeze feeder assemblies, 39, 77
Architectural plans, 19–20
Atmospheric burners, 69, 71
Automatic dampers, 158, 159
Axial fans, 149

Balancing cost estimates, 47, 190–197
Baseboard radiation, 92–93
Baseboard radiation controls, 172
Base-mounted pumps, 73, 76
Bathroom fixtures, 62, 64–66
(*See also* Toilet exhaust ductwork; Toilet exhaust fans)
Bathtub assemblies, 65
Block load ductwork, 158
Boiler assemblies, 36
Boiler controls, 170
Boiler heating plants, 68–69, 71–72
Boiler loads, 25
Boiler room piping, 72
Boiler standards, 69, 71
Boiler water feeders, 71–72
Booster pumps, 72–73, 75
Booster systems, 55, 57, 58
Breeching to flues, 71
Building components in cost estimating, 3–4
(*See also specific entries, for example*: Cooling systems; Heating, ventilating, and air conditioning systems)
Building cost adjustment indices, 6, 9–18
Building loads, 26–27, 31
Buildings:
construction cost escalation for, 6–8

Buildings (*Cont.*):
  heat losses of, 84–86
Built-up filter banks, 151, 153
Burner standards, 69, 71
Bypass decks, 177

Cabinet fans, 149
Cabinet heater assemblies, 82–83
Cabinet heater controls, 172
Cabinet heaters, electric, 91–92
Cabinets, fire hose, 205, 206
Cast iron boilers, 69, 71
Ceiling panels, radiant, 88–90, 93–94
Centrifugal fans, 147–148
Centrifugal water-cooled chillers, 96–98,
  106, 107
Chemical treatment assemblies, 103–104
Chilled water cooling systems, 34–35
  (*See also* Coolers; Cooling; *entries be-
    ginning with the term*: Cooling)
Chilled water mains:
  piping, 105–106
  transmission, 41
  (*See also* Cold-water mains)
Chilled water plants, 106–107
Chilled water pumps, 38
Chiller/condensers, 174
Chiller plant controls, 172–174
Chiller plants, 38, 96–98, 106–107, 174
Chiller room piping, 99
Chiller standards, 98–99
Chillers, 35, 96–98, 107
Cleanouts, 24, 53–54
Coil assemblies:
  fan, 107–110
  hot-water reheat, 82
  piping, 38, 41, 120–133
Coil banks, field-erected, 126, 127
Coil controls, reheat, 187
Coil loads, 31
Coil modules, 125
Coil piping equipment, 114
  (*See also* Coil assemblies, piping)
Coils:
  for air conditioning systems and heat-
    ing, ventilating, and air condition-
    ing systems, 124
  for furnace cooling equipment, 134–
    135, 137
  for heating and ventilating units, 123
  types of (*see specific entries, for exam-
    ple*: Cooling coils and pumps; Re-
    heat coils)

Cold decks, 181
Cold-water mains, 29–30
  (*See also* Chilled water mains)
Combustion air heater assemblies, 36, 72
Combustion air louvers, 71
Combustion air preheater controls, 170
Commercial-institutional radiation, 93
Compressed air stations, 169
Condenser assemblies, 99–100
Condenser chillers, 96–98, 107
Condenser water piping mains, 105–106
Condenserless chiller plants, 174
Condenserless chillers, 96–98, 107
Condensers, air-cooled, 174
Condensing controls, 174
Condensing units and coils, 134–135, 137
Connectors, flexible, 127
Control units, variable-air-volume, 44–
  45, 160–162, 181–188
Control valves, pneumatic, 169
Controls and instrumentation:
  cost estimating summary sheets for,
    189
  costs of, 167–189
  (*See also specific entries, for example*:
    Boiler controls; Temperature con-
    trols)
Convector assemblies, 83–84
Convector controls, 172
Cooler assemblies, 61
Coolers, 143–146
Cooling:
  for four- and six-row coils, 122
  pumped coil piping assemblies for, 130
Cooling assemblies, 213–214
  coil and coil piping, 41
  (*See also* Cooling tower assemblies)
Cooling coils and pumps, 35
Cooling equipment, 96–117, 134–135,
  137, 145
  cost estimating summary sheets for,
    42, 133, 145
  and piping, 42
  (*See also* Chillers; Cold-water mains;
    *entries beginning with the terms*:
    Air conditioning; Chilled water;
    Chiller)
Cooling generation and distribution, 40–
  42, 106
Cooling generation plants, 34–35
Cooling loads, 20, 26, 27, 113
Cooling pumps, 35, 172
Cooling systems, 34–35

Cooling tower assemblies, 100–103
 chemical treatment, 103–104
 freeze protection, 104–105
Cooling tower freeze protection controls,
 173–174
Cooling towers, 172–173
Cooling units, rooftop, 136–143
Cost adjustment indices, 6, 9–18
Cost escalation, 6–8
Cost estimates, balancing, 47
Cost estimating:
 principles of, 1–4
 starting, 19–21
Cost estimating summary sheets, 47–50,
 215–218
 for air distribution equipment, 46, 165
 for air moving equipment, 43, 164
 for controls and instrumentation, 189
 for cooling equipment, 145
 for cooling generation and distribution,
 40, 106
 for cooling systems, 42, 133
 for fire protection, 209
 for heating, 39, 40, 42, 90, 133
 for heating, ventilating, and air conditioning systems, 42, 133, 145
 for heating equipment, 70, 145
 for insulation, 201
 mechanical contractor's markups in,
 48, 49, 218
 multipliers in, 47–49, 215–218
 for plumbing, 32, 56
 for site work, 28, 52
Counterbalanced dampers, 158, 159

Damper controls, 170
Dampers, 158–160
Dates in cost estimating, 2
Diffusers, 45, 162–166
Distribution mains for perimeter heating
 systems, 36–37, 80–81
Dome-type exhaust fans, 149–151
Double-duct air conditioning and heating
 and ventilating units, 120
Double-suction base-mounted pumps, 73,
 76
Drains, plumbing, 61–67
Drinking fountain assemblies, 61
Dry pipe sprinkler valve assemblies,
 207–208
Ducts, 154–166
 balancing, costs of, 192–197

Ducts (*Cont.*):
 insulation of, 200
 types of: air, 44, 157–158
 block load, 158
 double, 120
 dual, 185–188
 galvanized steel, 157
 return, 155–157
 room load, 158
 round, 157
 supply, 44, 155, 156
 toilet exhaust, 43, 154, 158
 variable-air-volume, 44

EAT (entering air temperature), 82
Economic conditions in cost estimating, 3
Electric heaters, 37, 89–94
 baseboard hot-water radiation, 92–93
 cabinet, 91–92
 unit, 89–91
Electric heating, cost estimating summa-
 ry sheets for, 40, 90
Electric water cooler assemblies, 61
End-suction base-mounted pumps, 73, 76
Entering air temperature (EAT), 82
Entering water temperature (EWT), 82
Entry heating, 34
Equipment:
 balancing, cost of, 192–197
 capacity of, 3
 insulation of, 200
 selecting, 21
 types of (*see specific entries, for exam-
 ple*: Air distribution equipment;
 Plumbing equipment, costs of)
Equivalents, table of, 220
Evaporative coolers, 143–146
Evaporative cooling assemblies, 213–214
EWT (entering water temperature), 82
Exhaust ductwork, 43, 154, 158
Exhaust fans, 42–43, 149–153, 188
Exhaust hoods and louvers, 158, 159
Exhaust registers, 45, 164
Expansion tanks, 98
 and air fittings, 71

Fan coil assemblies, 107–110
Fan coil piping equipment, 114
Fan coils, perimeter piping mains for,
 112–117
Fan-powered variable-air-volume units,
 187–188

Fan system controls, 181–187
Fans, 147–153
  axial, 149
  cabinet, 149
  centrifugal, 147–148
  exhaust, 42–43, 149–153, 188
  propeller, 151, 152
  residential, 151, 152
  return air, 181–185
  utility, 150
  vaneaxial, 149
  (*See also* Air moving equipment; Heating, ventilating, and air conditioning systems; Heating, ventilating, and air conditioning unit controls; Heating and ventilating unit controls; Heating and ventilating units)
Field-erected coil banks, 126, 127
Fire dampers, 159, 160
Fire entry stations, 203–205
Fire hose cabinets, 205, 206
Fire protection systems, 21, 202–209
Fire pump assemblies, 206–207
Flanged construction, 129
Flexible connectors, 127
Floor heaters, 94
Floor plans, 23–24
Flues, breeching to, 71
Forced-draft burner boilers, 71
Fountain assemblies, 61, 63
Freeze detection thermostat controls, 175
Freeze protection assemblies, 104–105
Freeze protection controls, 173–174
Fuel oil tank assemblies, 210–212
Furnace cooling equipment, 134–135, 137
Furnace-type equipment, 134–138
  (*See also* Heaters; Heating; *entries beginning with the terms*: Heater; Heating)
Furnaces, 134

Galvanized steel ducts, 157
Gas-fired water boiler heating plants, 68–69, 71–72
Gas piping, 71
Gas service, 25–28, 54
Gas unit heaters, 135–136, 138
Gas water heaters, 28, 30
Grilles, 45, 162–166
Grooved pipe fittings, 203

Halon fire protection systems, 208
Heat exchanger controls, 171
Heat losses of buildings, 84–86
Heater assemblies, 58–59, 72, 82–83
Heater controls, 172
Heaters:
  electric, 37, 89–94
    baseboard hot-water radiation, 92–93
    cabinet, 91–92
    unit, 89–91
    (*See also* Heating, types of, electric)
  floor, 94
  gas unit, 135–136, 138
  kickspace, 94
  wall, 94
  water, 58–59
    gas, 28, 30
  (*See also entries beginning with the terms*: Heater; Heating; Hot-water)
Heating, 33
  cost estimating for, 36–40
    summary sheets for, 39, 40, 90
  of entries, 34
  pumped coil piping assemblies for, 130
  types of: electric, 40, 90
    radiation, 84–90, 92–93
  (*See also* Heaters)
Heating, ventilating, and air conditioning (HVAC) systems:
  coils for, costs of, 124, 125
  cost estimates for, 3–4, 31, 33–50, 124
  cost estimating summary sheets for, 42, 133, 145
  equipment and piping for, 42
  load calculations for, 26–27, 31
  multizone units of, 125
  selection of, 19–21
  (*See also* Fans; *entries beginning with the terms*: Air conditioning; Fan; Heating and ventilating)
Heating, ventilating, and air conditioning (HVAC) unit controls, 178–181
Heating and ventilating (HV) unit controls, 175–177
Heating and ventilating (HV) units, 115–116, 118–133
  coil piping assemblies in, 120–133
  without coils, 116, 118
  costs for, 119, 123
  multizone and double-duct, 120
  (*See also* Fans; *entries beginning with the term*: Fan)

Heating assemblies:
    coil, 37–38
    piping, 38
    solar, 212–213
    (*See also* Hot-water assemblies)
Heating coil loads, 31
Heating coils, 33–34, 122–127
Heating controls, radiant, 187
Heating equipment, 68–96
    cost estimating summary sheets for,
        42, 70, 133, 145
    packaged, 134–138, 145
    and piping, 42
Heating loads, 20, 25–27, 31
Heating piping mains, 77–80
Heating plants, boiler, 68–69, 71–72
Heating pump assemblies, 36
Heating pumps, 31, 33, 172
Heating side fan controls, 185–187
Heating systems, perimeter, 36–37, 80–81
    (*See also* Heating, ventilating, and air
        conditioning systems)
Heating terminal units, 94–96
Heating units, packaged, 134–138
Hoods, exhaust and intake, 158–159
Horizontal fan coil assemblies, 107–110
Horizontal piping, 78–80
Horizontal unit ventilator assemblies,
    111
Hose cabinets, 205, 206
Hot decks, 177, 180–181
Hot-water assemblies:
    convector, 83–84
    generator, 59–60
    radiant ceiling panel, 88–90
    reheat coil, 82
    unit heater, 82–83
    (*See also* Heating assemblies)
Hot-water controls, 172
Hot-water loads, 20, 21, 25
Hot-water radiant ceiling panels, 93–94
    (*See also* Hot-water assemblies, radiant
        ceiling panel)
Hot-water radiation, baseboard, 92–93
Hot-water unit heater controls, 172
    (*See also* Hot-water assemblies, unit
        heater)
HV (*see* Heating and ventilating unit
    controls; Heating and ventilating
    units)

HVAC (*see* Heating, ventilating, and air
    conditioning systems; Heating,
    ventilating, and air conditioning
    unit controls)

Indirect-direct evaporative cooling assem-
    blies, 213–214
Indoor cooling towers, 173
Inflation rates, 2
In-line pumps, 73, 75, 76
Instrumentation (*see* Control units, vari-
    able-air-volume; Control valves,
    pneumatic; Controls and instru-
    mentation)
Insulation cost estimates, 47, 198–201
    summary sheets for, 201
Intake hoods, 44
    and louvers, 158, 159

Isolation bases, 98

Kickspace heaters, 94
Kitchen sink assemblies, 63

Labor in building cost adjustment indi-
    ces, 6, 9–18
Labor correction factors, 5–6
Labor rates, 4, 5
Lavatory assemblies, 62
Leaving air temperature (LAT), 82
Life safety requirements, 21
Line items, 2
Loads (*see specific entries, for example*:
    Cooling loads; Heating loads)
Location:
    building cost adjustment indices by, 6,
        9–18
    in cost estimating, 2, 20
Louvers, 71, 158, 159
Low-contour exhaust fans, 149–150

Main air lines, 169
Mains (*see specific entries, for example*:
    Piping mains; Water mains)
Make-up air units, 146
Manhole costs, 53–54
Manual opposed blade dampers, 158, 159

Materials in building cost adjustment indices, 6, 9–18
Mechanical contractor's markups, 48, 49, 218
Mechanical legend, 221
Motor costs, 210, 211
Multiple coil piping assemblies, 130
Multipliers in cost estimating summary sheets, 47–49, 215–218
Multizone unit controls, 176–177, 180–181
Multizone units, 120, 125

Oil tank assemblies, 210–212
Outdoor air ducts, 44, 157–158
Outdoor cooling towers, 173

Packaged cooling equipment, 136–143, 145, 174
Packaged heating equipment, 134–138, 145
Parallel configuration, 187–188
Penthouse plans, 23
Perimeter heating systems, 36–37, 80–81
Perimeter piping mains, 112–117
Pipe fittings, grooved, 203
Pipe insulation costs, 200
Pipe sprinkler valve assemblies, 207–208
Piping, 42
  boiler room, 72
  chiller room, 99
  gas, 71
  horizontal, 78–80
  vertical, 80
Piping assemblies:
  bridle, 81
  coil, 38, 41, 120–133
Piping costs, 53, 67, 200
Piping equipment, sizing, 114, 115
Piping mains, 67, 77–80, 105–106, 112–117
Piping systems:
  balancing, cost of, 192–197
  for fire protection, 202–204
  water, 105–106
Plenums, prefabricated, 151, 153
Plumbing equipment, costs of, 55–61
Plumbing estimates, 28–32, 56
Plumbing fixtures, 21
  and drains, 61–67
Pneumatic control valves, 169

Pneumatic tubing, 170
Pot feeder assemblies, 36, 76
Prefabricated plenums, 151, 153
Pressure-reducing valves (PRVs), 55
Primary-secondary piping bridle assemblies, 81
Propeller fans, 151, 152
PRVs (pressure-reducing valves), 55
Pump assemblies, 72–76, 206–207
Pump connectors, 38, 41
Pump controls, 172
Pumped coil piping assemblies, 130
Pumps:
  base-mounted, 73, 76
  booster, 72–73, 75
  chilled water, 38
  cooling, 35, 172
  heating, 31, 33, 36, 172
  in-line, 73, 75, 76
  sewage, 60, 61

Radiant ceiling panel assemblies, 88–90
Radiant ceiling panels, 93–94
Radiant heating controls, 187
Radiant panel controls, 172
Radiation assemblies, 37, 86–88
Radiation capacity, 33
Radiation controls, 172, 187
Radiation heating, 84–90, 92–93
Rangehood fire protection systems, 208–209
Reciprocating chillers, 35, 96–98, 107
Recirculating pump assemblies, 60
Registers, 45, 162–166
Reheat coil assemblies, 82
Reheat coil controls, 187
Reheat coils, 187–188
Relief hoods, 44
Residential fans, 151, 152
Residential radiation, 93
Return air ducts, 44
Return air fans, 181–185
Return air grilles, 45, 164
Return ductwork, 155–157
Roof exhaust fans, 149–150
Rooftop cooling units, 136–143
Room load ductwork, 158
Round ductwork and accessories, 157

Sanitary sewers, 52, 53
Scheduled supply fluid temperature controls, 171

Scotch Marine fire tube boilers, 71
Screwed fittings, 128
Series 60 in-line pumps, 73, 75
Series 80 in-line pumps, 76
Service sink assemblies, 62
Sewage pump assemblies, 60, 61
Sewer mains, 30
Sewer piping, 53
Sewers, 20, 22, 24
  sanitary, 52, 53
  storm, 20, 24, 52, 53
  (See also Cleanouts)
Shower assemblies, 65, 66
Sink assemblies, 62, 63
Site plans, 20, 25
Site work estimates, 22–28, 51–54
Size versus unit cost, 4
Slinger-type coolers, 145–146
Small hot-water reheat coil assemblies, 82
Smoke detection controls, 175
Solar heating assemblies, 212–213
Sound attenuators, 43, 151, 153
Specifications in cost estimating, 2–3
Sprinkler and piping systems, 202–204
Sprinkler valve assemblies, 207–208
Standpipes in fire protection, 205, 206
Starter costs, 210, 211
Static pressure sensors, 183–184
Steam-to-fluid heat exchanger controls, 171
Storm sewers, 20, 24, 52, 53
  (See also Cleanouts)
Street cut costs, 51
Sump pump assemblies, 60, 61
Supply ducts, 44, 155, 156
Supply fluid temperature controls, 171
System values, verifiable, 1

Task difficulty in cost estimating, 4
Teflon pump connectors, 38, 41
Temperature of air and water, 82
Temperature controls, 21, 45–46, 171
Thermostat controls, 175
Timetables in cost estimating, 20
Toilet assemblies, 64, 65
Toilet exhaust ductwork, 43, 154, 158
Toilet exhaust fans, 42–43
Transmission mains, 37
Travel time in cost estimating, 4
Tubing, pneumatic, 170

Two-deck multizone heating, ventilating, and air conditioning unit controls, 180–181

Unit cost versus size, 4
Unit heaters:
  electric, 89–91
  gas, 135–136, 138
Unit ventilator assemblies, 110–111
Unit ventilator piping equipment, 115
Unit ventilators, 112–117
Upblast exhaust fans, 150, 151
Urinal assemblies, 64
Utility fans, 150

Vaneaxial fans, 149
Variable-air-volume (VAV) air systems, 33
Variable-air-volume (VAV) control units, 44–45, 160–162, 181–188
Variable-air-volume (VAV) ductwork, 44
Ventilating (see Fans; Heating, ventilating, and air conditioning systems; Heating, ventilating, and air conditioning unit controls; Heating and ventilating unit controls; Heating and ventilating units; entries beginning with the term: Fan)
Ventilator assemblies, 110–111
Ventilator piping equipment, 115
Ventilators, 112–117
Vertical fan coil assemblies, 107–110
Vertical piping, 80
Vertical unit ventilator assemblies, 111

Wall heaters, 94
Wash fountain assemblies, 63
Water:
  quantities of, 20
  temperature of, 82
  (See also Chillers; Cold-water mains; Cooler assemblies; Coolers; Cooling; Heaters; Heating; entries beginning with the terms: Chilled water; Chiller; Cooling; Hot-water)
Water boiler heating plants, 68–69, 71–72
Water closet assemblies, 64, 65
  (See also Toilet exhaust ductwork; Toilet exhaust fans)

Water-cooled chillers, 96–98, 106, 107
Water cooler assemblies, 61
Water feeder assemblies, 99
Water feeders, 71–72
Water heater assemblies, 58–59
Water heaters, 28, 30, 58–59
  (*See also* Heaters; Heating; *entries beginning with the terms*: Heater; Heating; Hot-water)
Water mains, 29–31
  piping, 105–106
  transmission, 41
Water piping systems, 105–106
Water plants, chilled, 106–107
Water-pressure-booster systems, 55, 57, 58

Water-pressure-reducing-valve assemblies, 55, 57
Water pumps, chilled, 38
Water service, 20, 22, 51–53
  (*See also* Plumbing equipment; Plumbing estimates; Plumbing fixtures)
Water-to-fluid heat exchanger controls, 171
Weather conditions, 3–4
Welded and flanged construction, 129
Wet sprinkler and piping systems, 202–204
Wetted-pad-type coolers, 144

Zone control valve assemblies, 205

## ABOUT THE AUTHOR

James H. Konkel founded the highly successful engineering firm, McFall-Konkel & Kimball, twenty-five years ago. He received the ASHRAE Fellow Award and the ASHRAE Energy Conservation Award, was recently elected a Fellow in the American Consulting Engineers Council, and is a member of several professional societies, including the American Society of Mechanical Engineers. He has been a guest lecturer at the University of Colorado.